Geograph[y]

GCSE

Olly Phillipson and Vincent Bunce

WHAT THIS BOOK CONTAINS

How to do Rapid Revision

The aim of *Rapid Revision* is simple. It's to help you do better in your exams in a very short time. This book is for active revision – it puts you in charge. In order to do well in your exams you need to:

- have a plan of what you need to know – that's the Action plan on page 4
- quickly target what you need to learn – that's in the Revision Units which are at the heart of this book
- be able to learn quickly and effectively
- understand what the examiners are looking for – that's in the grade-boosting tips in the answers.

A few other things you need to know

There are eleven different geography GCSE syllabuses, so it is really important that you know which syllabus and which exam board you are studying – ask your teacher. Although the majority of syllabuses cover most of the topics in this book, there are differences. Check which of the topics are covered by your particular GCSE syllabus – whether they are equally important, compulsory or optional. This is essential before you can plan your revision -don't revise topics your exam doesn't cover.

You will be entered at either higher or foundation level. Higher papers cover grades A*, A, B, C and D. Foundation papers cover grades C, D, E , F and G. Both levels have two separate written exam papers which count for between 75% and 80% of your final mark (the remainder is coursework, not covered by this book). Make sure you know which topics are covered on each of the two papers so that you can use this book effectively.

This book can be used by all students to improve their final grades, right up to A*. It will prove especially useful for candidates looking to turn a D grade into a C grade, whether at higher or foundation level entry. However, all GCSE candidates can benefit from using this book as a guide to the topics covered at GCSE and as a basis for exam revision.

A special feature of the book is the section on Case Studies within each unit. These case studies cover both MEDCs (More Economically Developed Countries) and LEDCs (Less Economically Developed Countries). Some syllabuses have case studies which you have to study, but the majority allow teachers to choose. Most of the commonly used case studies are listed for you, with one or two in each unit described in more detail. These can be used to help you answer the third Exam Practice question at the end of each unit which is always based on a case study. However, for the final exams you should revise the actual case studies you have been taught.

How to get going

Step 1

Look at the action plan on page 4. Check on what you need to cover. You can start with the first unit on the list or target something that worries you.

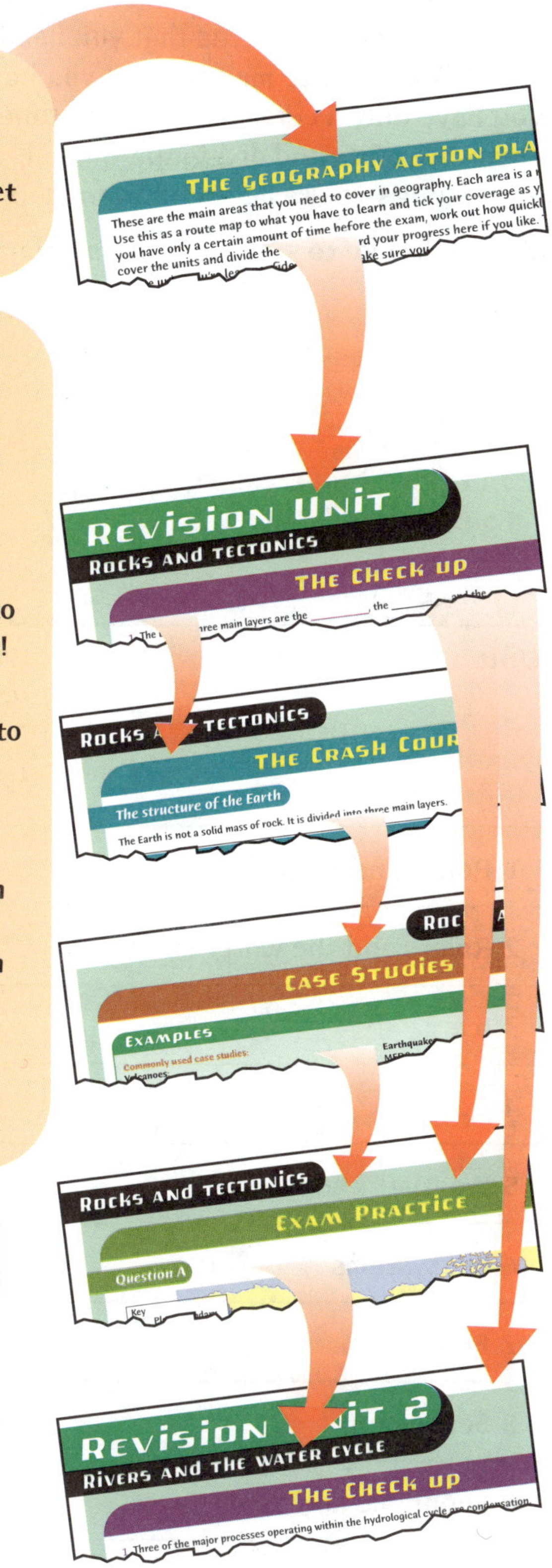

Step 2

Choose a Revision Unit. The Revision Units are in four parts:

1 The Check up

How well do you know the topic? Do this quick check to find out. If it seems OK you can skip the crash course and go straight to the exam practice - or you can go straight on to another unit. Rapid revision means targeting what you need to learn. It's very easy to spend time on things you already know!

2 The Crash Course

That's the bit that teaches you the key skills you need to have to do well in the exam. It puts the essential skills very clearly and simply - and it's broken into small chunks for easy practice.

3 Case Studies

There is a list of the most commonly-used case studies in each unit. It is very important that you revise the case studies you have been taught. It is likely that several of these are found in this book.

4 Exam Practice

These provide GCSE type questions section by section. They are a real test of what you need to do in the exam.

The Revision Units are really flexible - if you want, just do The Crash Course to get familiar with what you have to do - or focus on Exam Practice nearer the exams.

Step 3

Record your progress on the action plan if you want. Then go on to the next unit.

And if you're really short of time ...

1. Don't panic.
2. Do The Check ups. Start with the topics you're less confident about. If you do OK, go straight to the next unit. If not, do The Crash Course.
3. Write the key things from The Crash Course you want to remember in note form on cards and pin them to your bedroom wall.
4. Don't try to learn anything new the night before the exam.

The geography action plan

These are the main areas that you need to cover in geography. Each area is a revision unit. Use this as a route map to what you have to learn and tick your coverage as you go along. If you have only a certain amount of time before the exam, work out how quickly you need to cover the units and divide the time up. Record your progress here if you like. Then concentrate on the units you're least confident about. Make sure you only revise the topics in your syllabus.

You might like to start your action plan by finding out and recording:

Dates of exams

Tier (level) you're entered for Foundation / Higher

Unit	How many times revised			Topic included in exam syllabus?	Check up score /20	Exam Practice Questions attempted		
	1	2	3			QA	QB	QC
1 Rocks and tectonics								
2 Rivers and the water cycle								
3 Ice								
4 Coasts								
5 Weather and climate								
6 Ecosystems								
7 Population								
8 Settlement								
9 Economic activity								
10 Resources and energy								
11 Development								
12 Leisure and tourism								
13 Map skills								

Areas in which you feel you need more practice

As you revise each unit, write down any parts you may want to go over again in a second or third revision.

Unit 1			
Unit 2			
Unit 3			
Unit 4			
Unit 5			
Unit 6			
Unit 7			
Unit 8			
Unit 9			
Unit 10			
Unit 11			
Unit 12			
Unit 13			

The secrets of success in geography

Here are some key points to help you be confident that you are properly prepared for your two final exam papers and are ready to succeed:

1 **Get organised**: know exactly which topics are on each exam paper and the type or style of questions you will be asked. How many questions must you answer? Will you have a choice, or are some questions compulsory? Look at past papers as a guide. Are you likely to have extra resources to look at as well as the exam paper, e.g. OS maps, resource booklets etc? Don't be surprised - be prepared!

2 **Divide up your time carefully**: before you sit your exam, be clear how many questions / sections you have to answer. Are they all worth the same amount of marks? If so, allow equal time - it is pointless answering two sections perfectly and not having time for the other two ... you can only get 50%! Many exams give you guidelines in the rubric (instructions) or on the exam paper itself. Always allow at least five minutes within your timings to read through the whole paper and any resources you may be given.

3 **Learn your case studies**: these are actual examples of places which illustrate geographical theory, e.g. how cities grow in LEDCs (Rio de Janeiro, Calcutta etc.). These often carry high marks in the exam. Use the Case Study sections in the book as a guide to the type of information you need to learn.

4 **Use geographical language / terms**: make a list of all the specific geographical terms used for each topic you cover (practise by using The Crash Course sections in this guide). For each word or term write - and learn - a short definition explaining what it means. In the exam, many of these will be used as part of the exam questions - in some cases you may be asked to explain exactly what the term means. Your word lists can also act as a starting point for revising each topic.

5 **Practise your skills**: good geographers need to have a wide range of skills. Make a list of all the different skills you have used throughout your GCSE course. Use headings like 'types of graphs', 'writing skills'. For example, types of graphs may include pie charts, divided bar graphs, line graphs, block / bar graphs and scattergraphs. Make sure you know how to draw / use / interpret each type.

6 **Dissect the question**: instructions on exam questions use a range of command words. These range from quite basic commands - read, study, shade in, complete etc. through to more complex commands - explain, account for, annotate, analyse. Questions using more complex command words are invariably worth more marks.

7 **Get the length of your answer right**: sometimes there will be spaces on the exam paper to write in your answer - sometimes you may be given a booklet which is just lined paper. If there is a space given, use this as the guide as to how long your answer should be. A more dependable guideline is to see how many marks are available for that particular question or part of question. Don't write half a page if there are just 1 or 2 marks available. You are wasting valuable time and won't get more than the maximum marks available.

8 **Getting stuck**: if you do get stuck, move on to the next question and if possible come back to the problem at the end. Be careful, though, if it is part of a question where the next part relies on an earlier answer. Exam questions are sometimes quite lengthy, asking for a number of fairly short answers, which gradually become more difficult.

9 Exams MAY seem to be very stressful - but if you have worked hard and revised properly, they are also a great chance to show what you know and understand. Show the examiner what a good geographer you are ... but read each question carefully, think before you answer it, then write clearly.

10 Good luck!

Revision Unit 1

Rocks and tectonics

The Check up

1 The Earth's three main layers are the __________, the __________ and the __________.

2 The middle layer, the __________, is __________ rock.

3 Tectonic __________ are large slabs of rock which make up the Earth's surface.

4 __________ rocks are formed from the cooling down of liquid magma.

5 Look at diagram (a). Complete the labels for A, B and C.

6 Earthquakes displace rocks along lines of weakness called __________.

7 The majority of the world's major earthquakes occur along __________ boundaries.

8 Most of the world's earthquakes or tremors are quite small and can only be detected by an instrument called a __________.

Diagram (a)

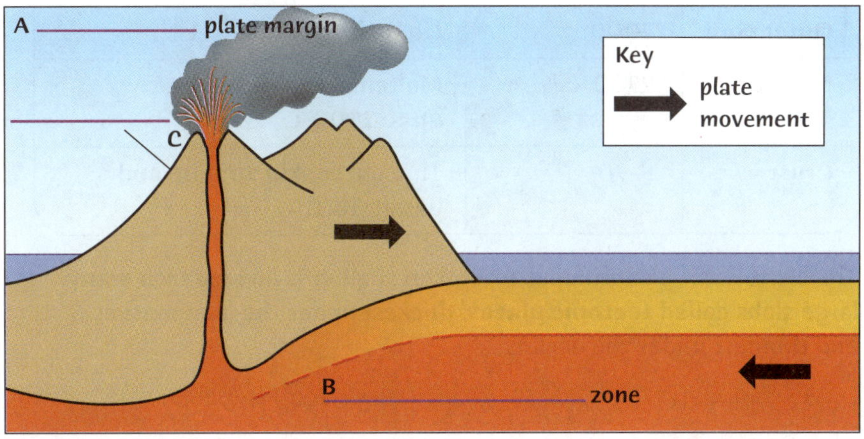

9 The strength of an earthquake is measured by the __________ Scale.

10 __________ can result from an earthquake which starts under the sea causing __________ to low-lying coastal areas.

11 Fires often break out after a major earthquake because __________.

12 __________ volcanoes are usually the most dangerous.

13 Ash, steam and other gases released when a volcano erupts often cause __________ flows.

14 A dormant volcano is one which __________.

15 Liquid rock is called __________ underground and __________ when it reaches the surface.

16 Look at diagram (b). Complete the labels.

17 Volcanoes are often found close to plate margins above __________ zones.

18 The circle of live volcanoes around the Pacific Ocean is called the __________.

19 __________ cools down to form new rock which often weathers down to give __________ soil for farming.

20 Ash released into the atmosphere from a major volcanic eruption can disrupt __________ patterns for many years.

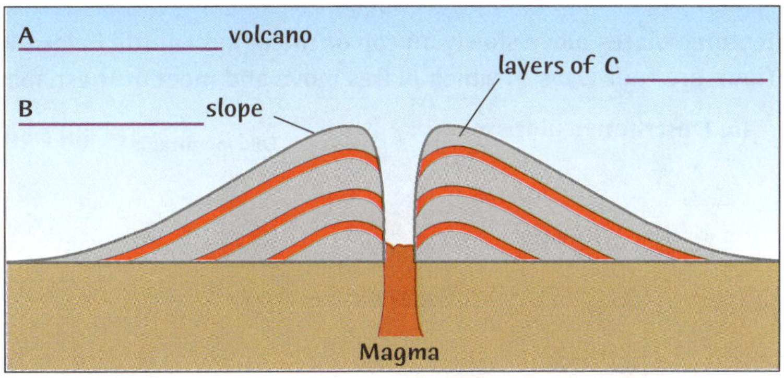

Diagram (b)

Mark yourself (see answers on page 96)

- More than 12? Do the Exam Practice – or go on to another unit.
- Less than 12? Do the Crash Course on the next page.

/20

The Crash Course

The structure of the Earth

The Earth is not a solid mass of rock. It is divided into three main layers.

Layer	Approx. thickness (km)	Description
Inner core	1400	Centre of Earth, solid metal iron and nickel, up to 3000°C
Outer core	2000	Liquid iron and nickel
Mantle	2900	Molten plastic-type rock, up to 2000°C
Crust	5-90	Thin outer skin, granite and basalt, 10°C

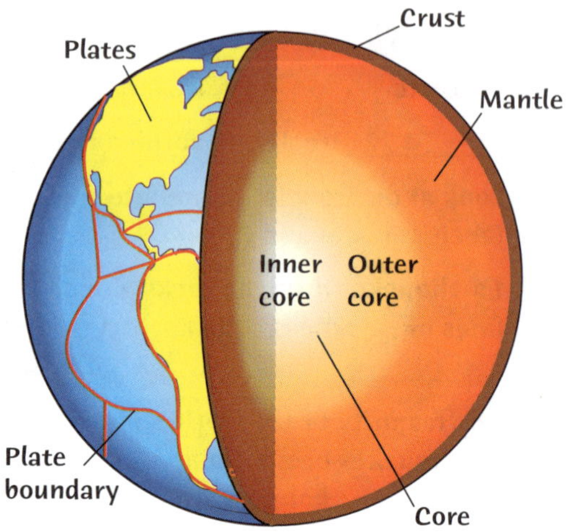

The crust is *not* a continuous unbroken shell. It is divided into many large slabs called **tectonic plates**, thickest under the land masses and thinnest under the oceans.

Rock types

Three types of rocks make up the plates which form the Earth's surface:

- **igneous**: formed when magma (lava) cools down;
- **sedimentary**: formed from the weathered or eroded particles of existing rocks;
- **metamorphic**: formed when existing rocks are changed by great pressure and / or heat.

Moving plates

Tectonic plates move slowly on top of the liquid mantle below. Where they meet is called a **plate margin**. There are four ways in which plates move and meet at these margins.

(a) Destructive plate margin

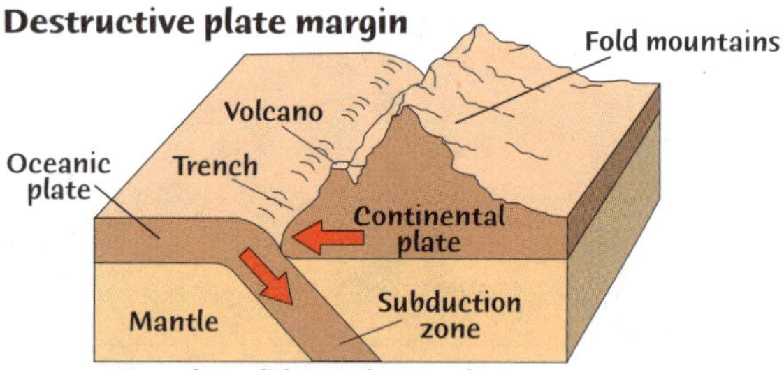

One plate slides under another

(b) Collision plate margin

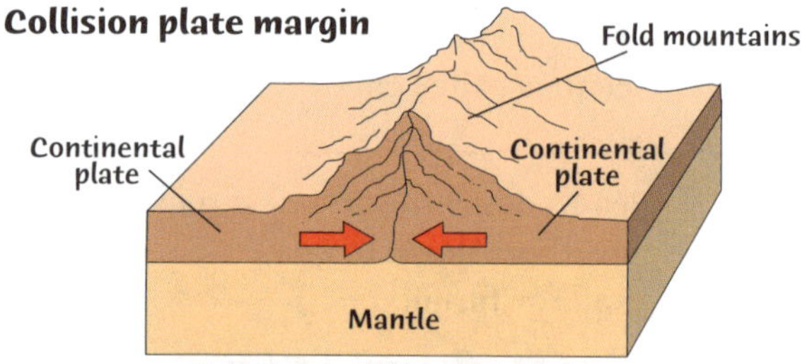

Two plates collide and push up

(c) Constructive plate margin

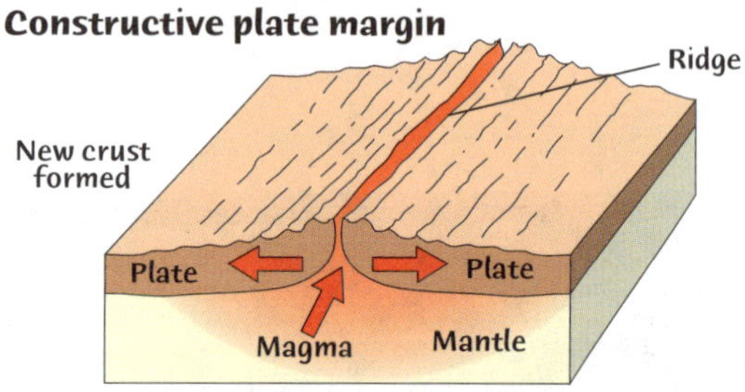

Two plates move apart creating new land

(d) Conservative plate margin

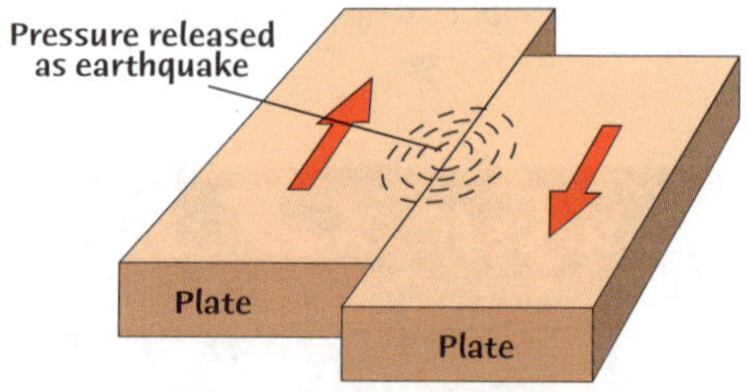

Two plates slide past each other

The Crash Course

Earthquakes

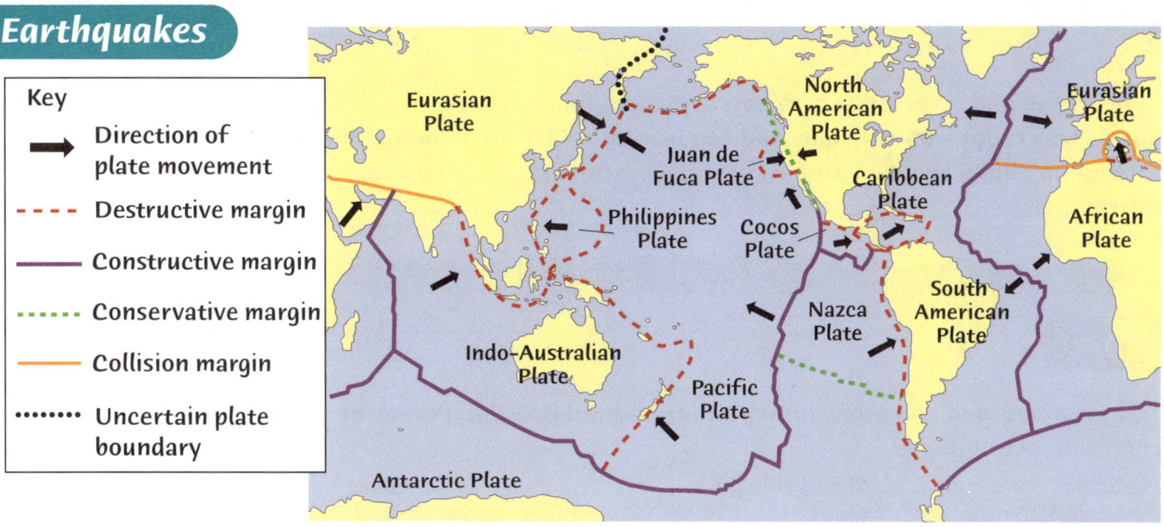

Distribution of plate margins and earthquake zones

Key facts to learn:

- Earthquakes usually result from the movement of tectonic plates. As plates move they create seismic waves, spreading out from the centre and eventually reaching the Earth's surface.
- Earthquakes displace rocks along lines of weakness in the Earth's crust called faults.
- Seismic waves come in batches – the main or **primary wave** is the biggest, followed by a series of smaller, **secondary waves**.
- Thousands of small, often almost undetectable, earthquakes or tremors happen every year.
- Larger earthquakes can cause immense damage, injury and loss of life.
- The instruments used to detect and measure earthquakes are called **seismographs**.
- The strength or intensity of an earthquakes is measured on the **Richter Scale**.
- An earthquake which starts under the sea bed can cause **tsunamis** – huge waves of up to 30 metres in height cause widespread damage and flooding along coastal areas.

Primary effects of a major earthquake

- The Earth shakes violently, causing buildings, roads and bridges to collapse, trapping, killing and injuring people.
- **Liquefaction** (the ground behaving more like liquid than solid) causes even earthquake-proof buildings to sink and collapse.
- Tsunamis cause huge waves which flood and destroy low-lying coastal areas.

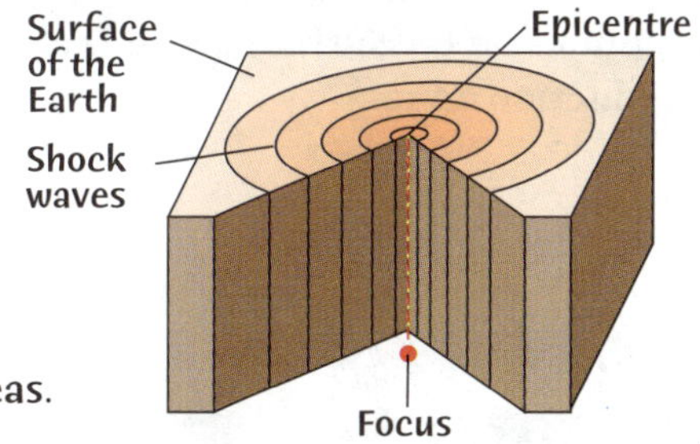

Secondary effects of major earthquakes

- Water and sewage pipes fracture, leading to water shortages, contamination and diseases.
- Fires break out as gas pipes are broken and gas ignites. With broken water pipes it is hard to put out these fires.
- Transport links are cut or blocked – people and emergency services cannot move about freely.
- Electricity and telephones may be cut off making communication and 'normal' life difficult.
- The economy may be badly affected (possibly for a long time) as industry cannot function.
- The cost of rebuilding is usually many millions of pounds – especially difficult for LEDCs.

The Crash Course

Volcanoes

When pressure building up under the crust is released, **magma** (molten rock) can escape to the surface as **lava**. Sometimes volcanoes may erupt very suddenly and explosively, releasing great quantities of ash, steam and lava. Alternatively, more liquid, runny lava may find its way slowly into cracks in existing rock or pour more gently through to the surface.

Types of volcano

A volcano may be active (erupted recently); dormant (not erupted for a long time); or extinct (no historical records of an eruption).

Composite volcano

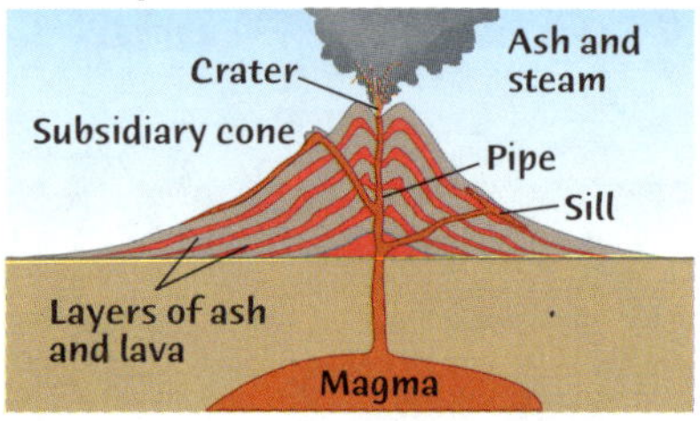

Shield volcano

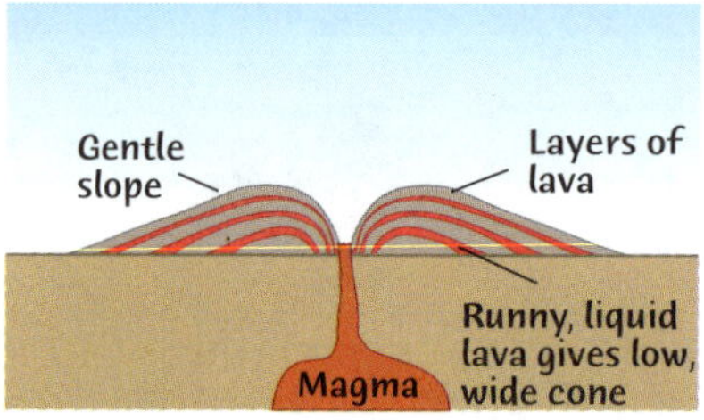

Caldera

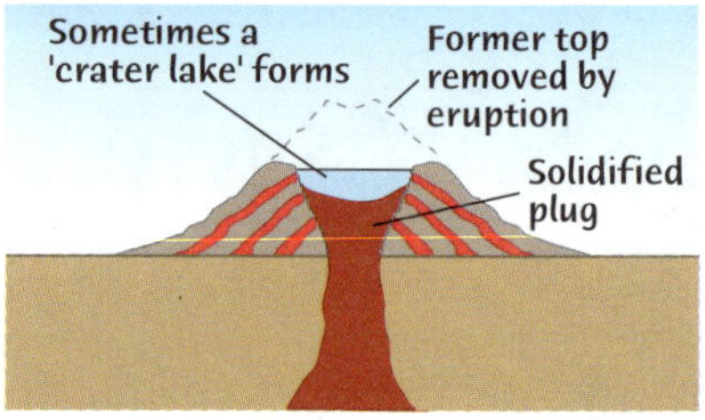

Distribution of the world's volcanoes

There are over 600 active volcanoes in the world, mostly found along or close to constructive and destructive plate margins (see p. 9). At constructive margins, plates move apart, allowing lava to pour gently to the surface. Along destructive margins crustal rock melts as one plate slides under another (subduction). This creates intense pressure, eventually leading to explosive eruptions on the surface above the **subduction zone**.

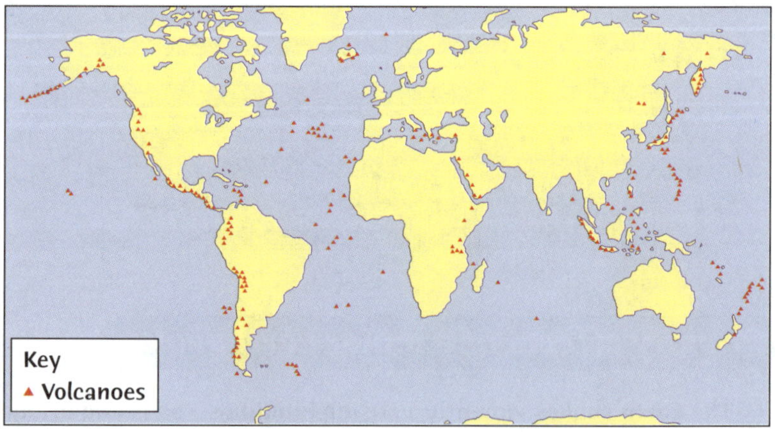

Distribution of the world's volcanoes

Effects of volcanic eruptions/activity

- Lava flows cover the land surface and set fire to buildings or vegetation.
- Ash, steam and other hot gases (called **pyroclastic flows**) can cause immense damage. Sometimes travelling at over 100mph, they can strip, burn and destroy vegetation and trees.
- Hot ash mixing with melting snow and ice on mountain tops or falling into rivers turns them to **mudflows** (lahars) which flow rapidly down hillsides and valleys.
- Ash falls to the ground covering everything, sometimes metres deep.
- New rock / land is created as lava cools down.
- As rock weathers it often creates rich, fertile soil for local farmers.
- Ash released into the atmosphere can affect or alter global weather patterns for many years.

Case Studies

Examples

Commonly used case studies:

Volcanoes:
MEDCs:
Mount St Helens, Washington State, USA (1980);
Mount Etna, Italy; Kilauea, Hawaii, USA (1983)
LEDCs:
Soufriere Hills, Montserrat, Caribbean (1995);
Mount Pinatubo, Philippines (1991)

Earthquakes:
MEDCs:
Kobe, Japan (1995); San Francisco, USA (1906, 1989);
Los Angeles, USA (1994); Assisi, Italy (1997)
LEDCs:
Mexico City, Mexico (1985); Maharashtra, India (1993);
Turkey (1999)

Sample case study: Kobe earthquake, Japan, 1995

Location:

The epicentre was under Awaji Island in Osaka Bay, close to the cities of Osaka and Kobe.

Description:

- main earthquake measured 7.2 on the Richter Scale and lasted 20 seconds
- many smaller aftershocks
- happened on 17 January 1995 at 5.46 a.m.
- over 5000 killed; 25 000 injured
- 300 000 made homeless; 200 000 buildings destroyed
- fires broke out as gas mains were fractured – these were spread by strong winds
- a major elevated expressway (motorway) collapsed as did several main railway lines
- fires and blocked roads made rescue attempts very difficult
- new Kansai airport unaffected
- main damage was in Kobe, with some damage in Osaka and Kyoto
- older buildings worst hit – newer 'earthquake-proof' buildings fared better
- water and food shortages followed
- total cost estimated at almost $100 billion

This earthquake showed that even a wealthy country like Japan can be severely affected by a major earthquake, both in the short term and in the long term. Japan lies along the meeting point of three of the world's main tectonic plates and experiences up to 7000 earthquakes per year. Fortunately very few are as strong as the one which hit Kobe.

Collapsed expressway in Kobe

Exam Practice

Question A

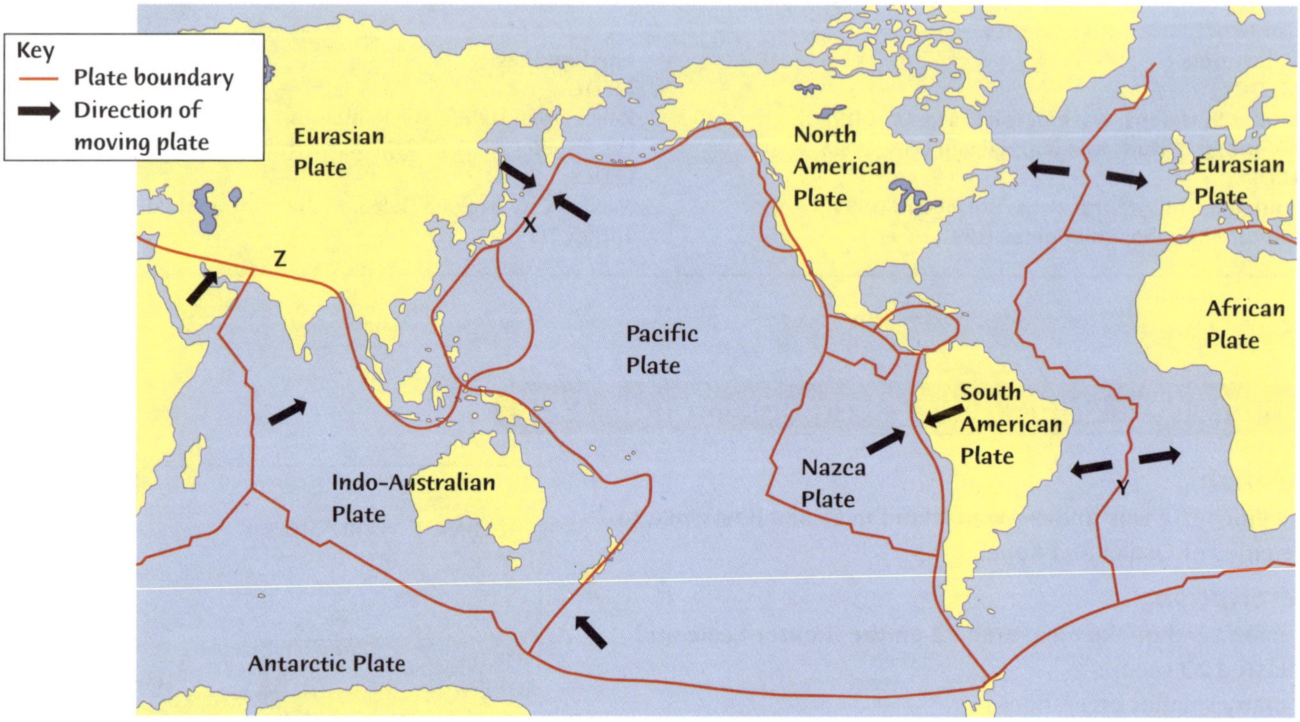

a What type of plate boundary is found at: (i) X (ii) Y (iii) Z?

b Draw an annotated diagram to show the type of plate boundary found at X.

c Which location – X, Y or Z – is most likely to have volcanic activity? Give reasons for your choice.

Question B

a Complete the labels on the diagram showing a composite volcano.

b Draw a labelled diagram of a shield volcano.

c Describe and explain three main differences between composite and shield volcanoes.

d Explain why many people choose to live close to active volcanoes despite the risk of eruptions.

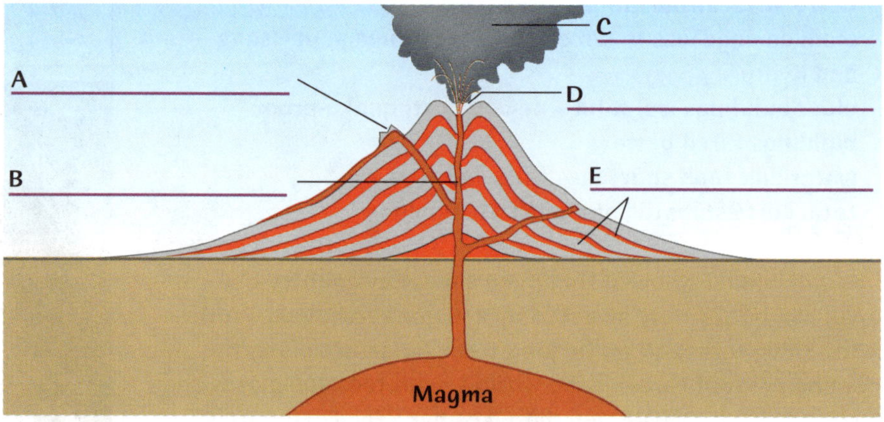

Question C: case study question

For a named earthquake you have studied:

a Describe the location of the earthquake. Why did it occur there?

b What were the main primary and secondary effects of the earthquake?

c i) What had been done to try to prevent earthquake damage?
 ii) What other measures could have been taken or could be taken in the future?

Revision Unit 2

Rivers and the water cycle

The Check up

1 Three of the major processes operating within the hydrological cycle are condensation, __________ and __________.

2 River or drainage basins, like most systems, consist of __________, outputs and __________.

3 The area where precipitation falls / water collects to feed a drainage basin is called the __________ area.

4 Water sinks into the soil by a process called __________.

5 __________ by trees and vegetation may delay rainfall reaching the ground.

6 Water moving or flowing through rocks underground is called __________ flow.

7 Eroded material carried by a river is called the __________.

8 The four main processes of river erosion are __________, abrasion, corrosion and __________.

9 Corrosion occurs when water __________ certain chemicals in the rocks forming the river channel.

10 The speed of river flow influences erosion and deposition. When the river flows slowly, more __________ occurs. When it flows quickly, __________ takes place.

11 The starting point of a river is called the __________. From here the river flows down to its __________ where it enters the sea.

12 Rivers are usually divided into three sections – the __________, __________ and __________ course.

13 Look at diagram (a) showing the formation of a waterfall. Complete the labels.

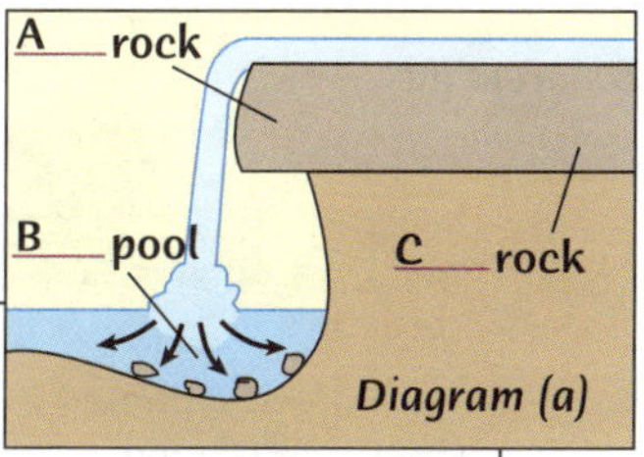

Diagram (a)

14 As a river begins to slow down (in its __________ course), it starts to form bends called __________. Eventually, the bends can become cut off from the main river channel, forming __________.

15 Look at diagram (b) – the last section of a river as it approaches the sea. Complete the labels.

Diagram (b)

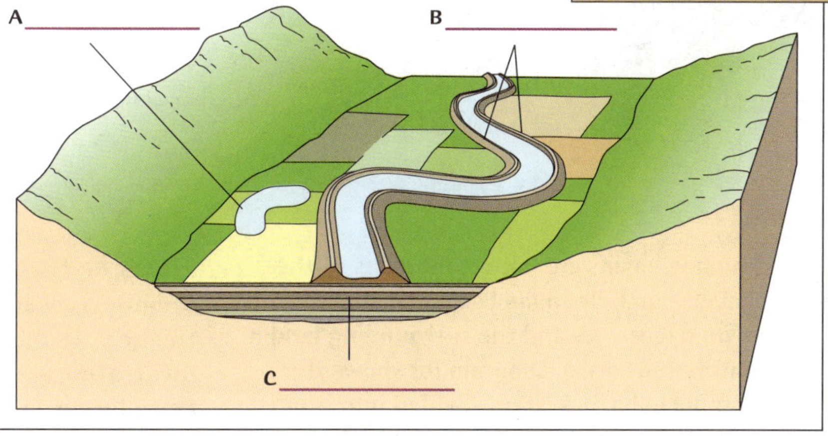

16 As deposition takes place on the river bed (see diagram b), the channel becomes __________, increasing the risk of __________.

17 Hydrographs show the amount of water or the __________ in a river at any given time.

18 __________ rainfall, __________ soil and lots of buildings and concrete speed up __________ run-off, allowing water to reach rivers more __________ and increasing the risk of flooding.

19 There is less risk of flooding when water reaches rivers more __________ e.g. because of __________ by vegetation, __________ rock and __________ built to control flow.

20 Despite the risks from flooding, flat, fertile lowland next to rivers (the __________ __________) is in great demand for farming, building and transport routes.

Mark yourself *(see answers on page 96)*

- More than 12? Do the Exam Practice – or go on to another unit.
- Less than 12? Do the Crash Course on the next page.

/20

THE CRASH COURSE

The hydrological cycle

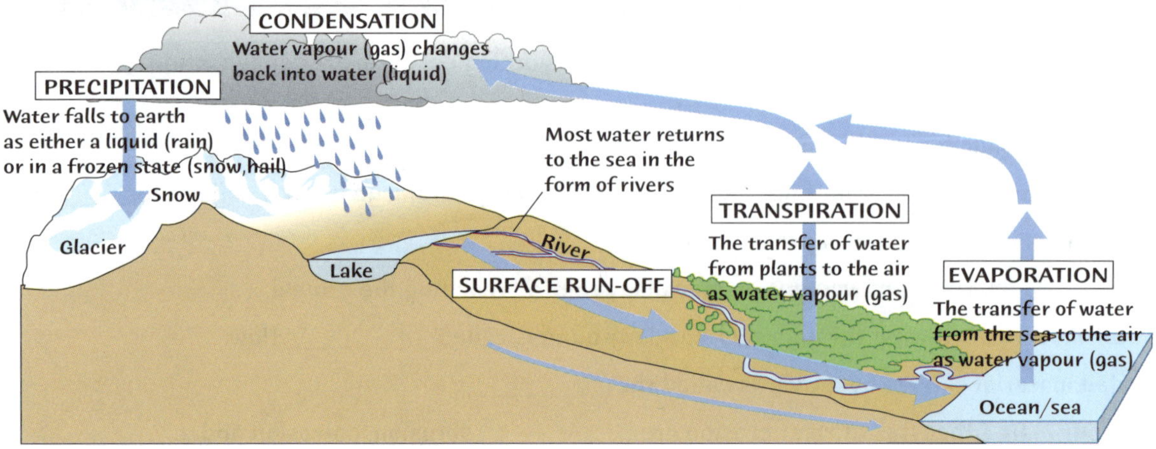

Diagram (a)

Hydrology is the study of water. The hydrological cycle in diagram (a) shows water being stored or moving around in its various states via five main processes – condensation, evaporation, precipitation, transpiration and surface run-off.

River/drainage basins

Diagram (b)

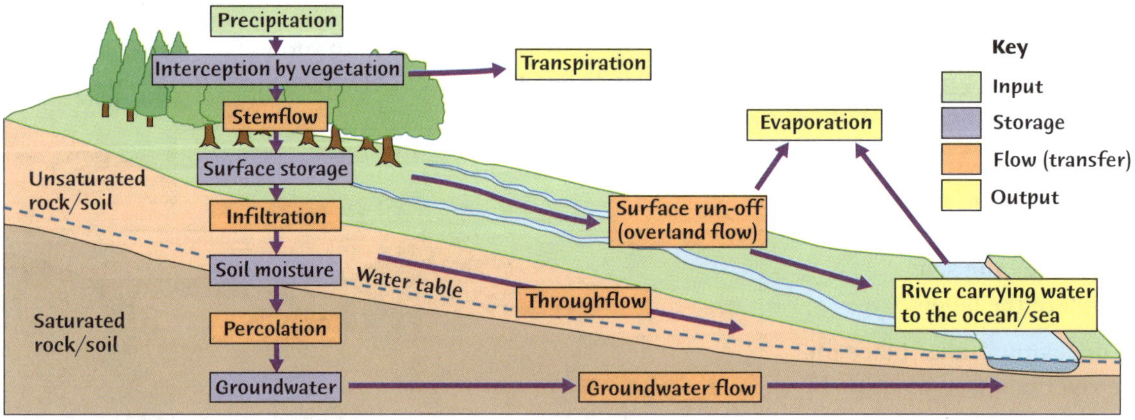

Individual drainage basins are important parts of the hydrological cycle. Each drainage basin is made up from a major river, its tributaries and the surrounding land it drains (the **catchment** area). Diagram (b) shows the parts of the drainage basin system, divided into **inputs**, **stores**, **flows** and **outputs**.

Inputs

Precipitation*

Stores

Interception – precipitation falls onto trees, plants etc. intercepting / delaying its progress to the ground
Surface storage – water remaining on the land surface e.g. lakes
Soil moisture – water stored in the soil layer
Groundwater – water stored deep underground in rocks

Flows or transfers

Stemflow – water dripping down the stems of plants, trees etc. to the ground after interception
Infiltration – water sinking into the soil
Percolation – water moving down from the soil into underlying rocks
Throughflow – water flowing through the sloping soil layer to a river
Groundwater flow – water flowing (slowly) through underground rocks
Surface run-off (overland flow)*

Ouputs

Transpiration*
Evaporation*
River – water flowing out of the basin to the sea / ocean

* see diagram (a)

The Crash Course

Processes

Apart from forming part of the hydrological cycle, rivers play a very important part in shaping the landscape. Rivers cut channels, then as they flow they erode the bed and banks of that channel. Rocks and soil which may fall or be washed into the river will also be eroded. Rivers then transport eroded material (called the **load**) before depositing it later.

A typical river in its upper course

Key facts to learn: the processes of erosion

- **Attrition**: pieces of material / rocks moving along the river bed are ground down into smaller fragments as they collide with one another
- **Abrasion / corrasion**: material carried by the river scrapes against the banks and bed, wearing it down
- **Corrosion**: water dissolves certain minerals or chemicals in the rocks of the banks and bed
- **Hydraulic action**: as water hits the river banks, air trapped and compressed in the rock can cause it to break and fall.

Transportation

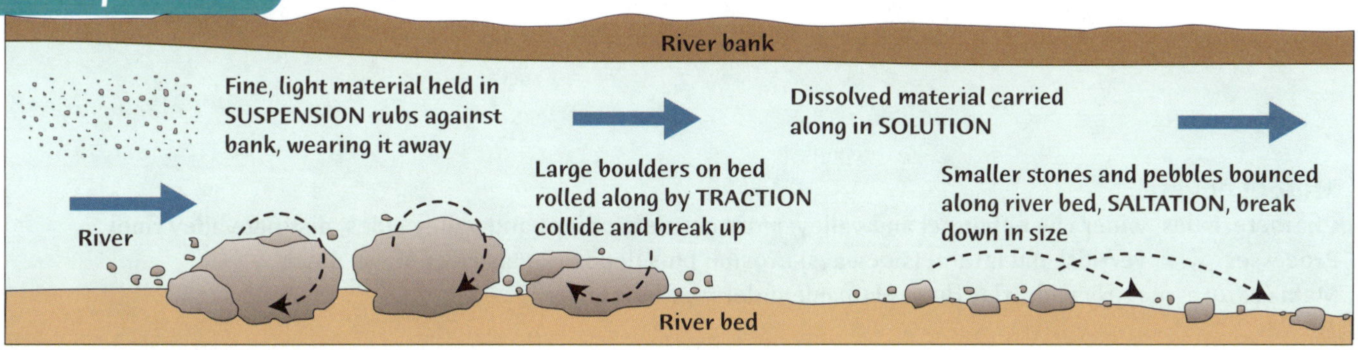

Deposition

When a river flows quickly it can erode and transport material.
When it slows down, especially in its lower course, material is deposited (dropped).

River profiles

A river runs from its source (usually in a highland area) to the sea (its mouth). Its course can be divided into three distinct sections or stages. As the slope (gradient) changes, the speed of the river slows down. The cross-section here shows the **long profile** of a river.

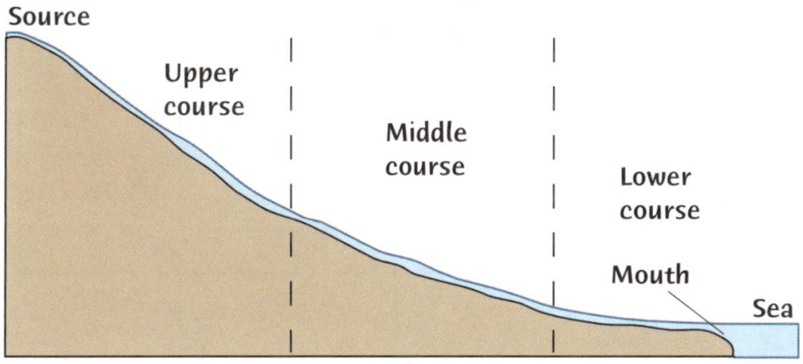

THE CRASH COURSE

River landforms

Upper course

Characteristics: steep gradient, narrow, fast-flowing, winding course
Processes: mainly vertical (downwards) erosion
Main features: V-shaped valley, interlocking spurs, potholes, rapids, waterfalls and gorges

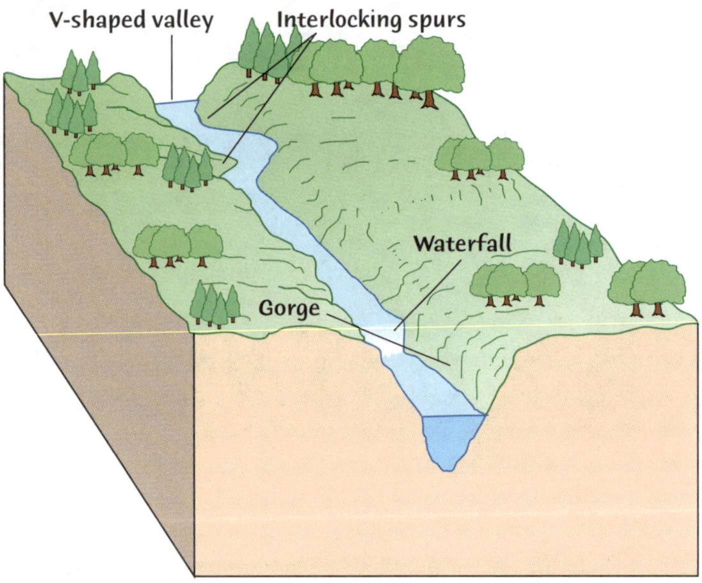

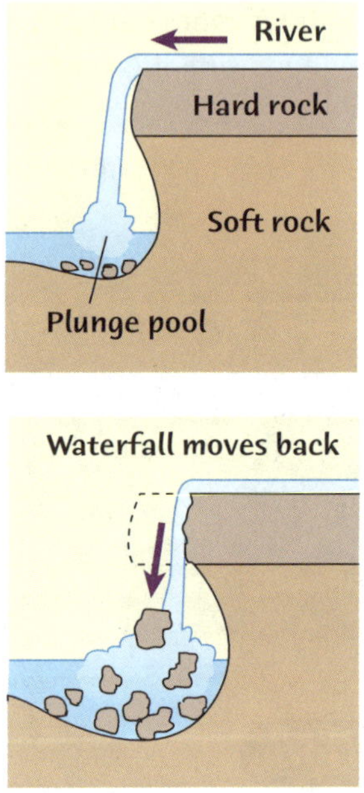

Middle course

Characteristics: wider river channel and valley, more gentle gradient and valley sides, distinct valley floor
Processes: vertical and lateral (sideways) erosion plus deposition as river slows down
Main features: fairly wide, flat flood plain, meanders, ox-bow lakes

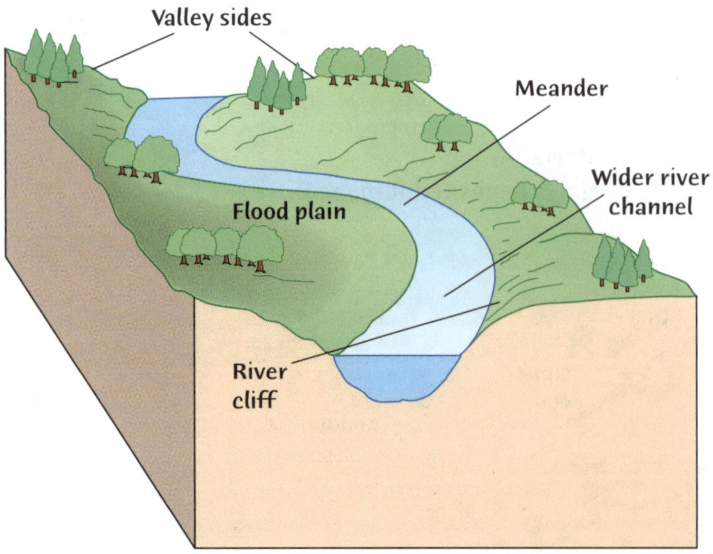

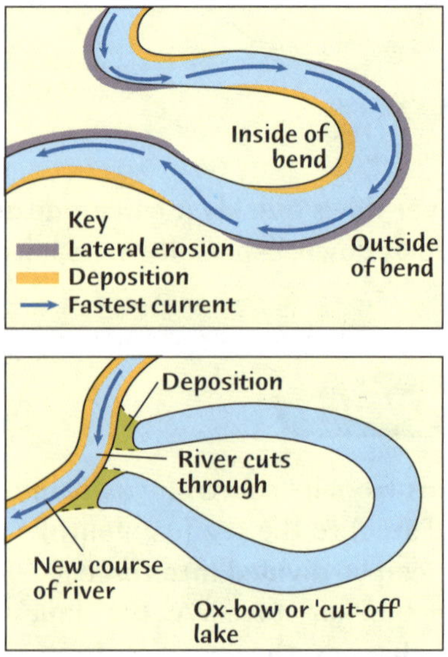

THE CRASH COURSE

Lower course

Characteristics: wide channel and almost flat, wide valley floor, tidal near mouth
Processes: little or no erosion, mainly deposition as river flow is very slow
Main features: embankments (levées), meanders, ox-bow lakes, estuary, mouth, delta, mudflats

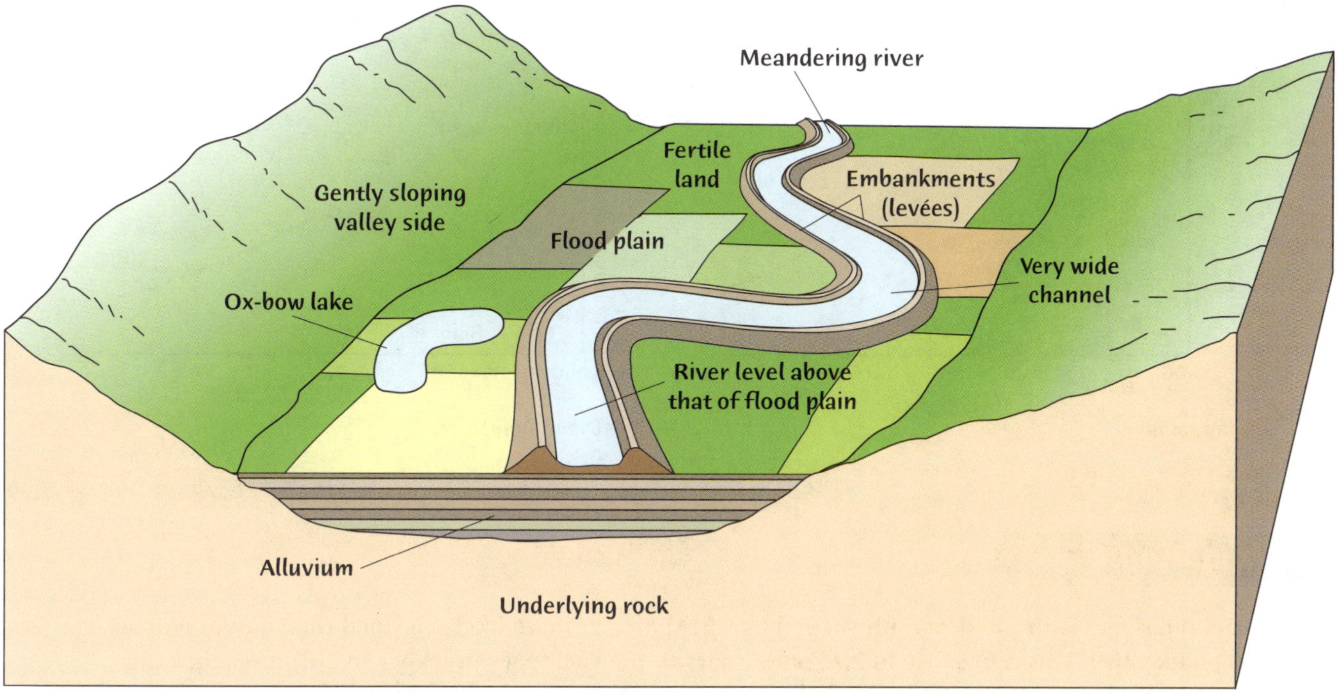

A typical view of a river's lower course as it flows into the sea. Note the wide channel and areas of deposition.

The Crash Course

Hydrographs

River engineers need to know which rivers in an area are likely to flood so that appropriate planning and management can take place to protect nearby land and prevent damage. **Flood hydrographs** help by measuring the amount of water (**discharge**) in a river at any given time. Flood hydrographs show how quickly or slowly river levels may rise after heavy rain.

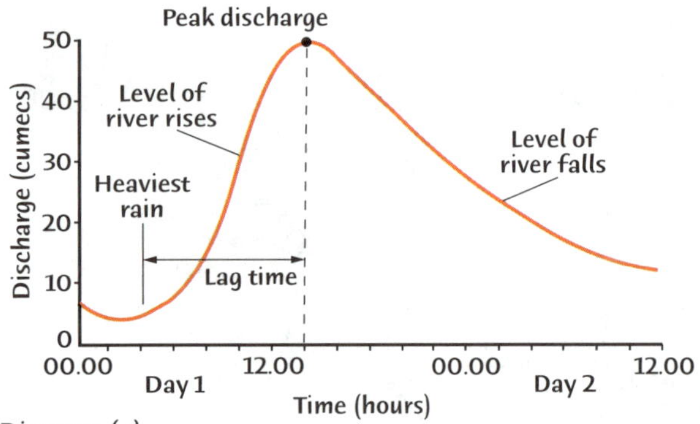

Diagram (a)

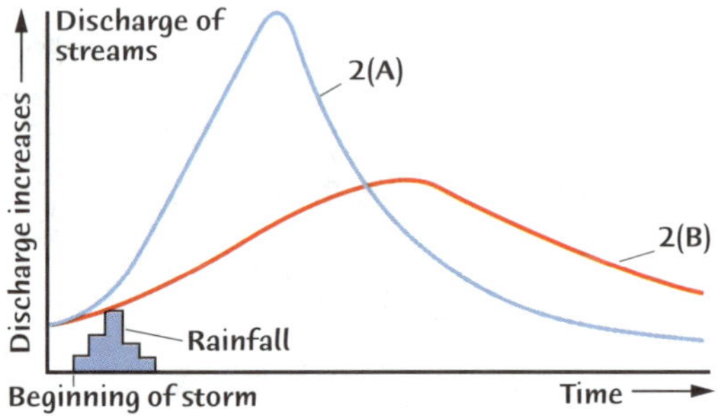

Diagram (b)

River management

In diagram (b), rivers with a hydrograph shaped like 2(A) are far more likely to flood than a river with a hydrograph like 2(B). This is because in 2(A) water gets to the river very quickly – in 2(B) it reaches the river more slowly and gradually. The table below shows some of the possible reasons for this.

	2(A) – hydrograph rises rapidly	2(B) – hydrograph rises gradually
Amount of rainfall	Heavy	Light
Type of rock	Impermeable	Permeable
Land use	Urban – towns / cities (concrete)	Rural – fields etc.
Vegetation	Bare soil	Forest
Management	No dams	Dams on river
Slope	Steep	Gentle / flat
Basin shape	Circular	Elongated (long/narrow)
Drainage density	High – more streams and tributaries	Low – few streams or tributaries
Climate / weather	Very cold / hot / wet: frozen, hard, sun-baked or saturated ground = high surface run-off	Cool / warm / dry: water sinks into ground = low surface run-off

Land next to rivers is often full of human activity – with housing, industry, farming and transport routes competing for space. This means that flooding can cause immense damage, injury or death. Most rivers are therefore managed or altered in a number of ways. However, human activity can actually make flooding more likely e.g.

- building on the flood plain can increase surface run-off
- clearing vegetation and trees stops interception which usually delays rainfall reaching a river
- preventing flooding e.g. by building embankments / walls, can cause a build-up of silt on the river bed, making the channel shallower
- management schemes on some parts of a river can cause problems in other sections e.g. if water behind a dam becomes too high, it has to be released downstream.

Case Studies

Examples

Commonly used case studies:

UK: Rivers Tees, Severn, Ure plus local studies

MEDCs:
River Mississippi (USA); River Colorado (USA): Rivers Rhône / Rhine (Europe)

LEDCs:
River Nile / Aswan dam (Egypt); River Amazon (Brazil); River Kano (Nigeria); River Yangtze / Three Gorges Dam (China)

Sample case study: Mississippi River Basin, USA

Location:

The Mississippi River stretches from the Canadian border in the north, 3800km south to its delta in the Gulf of Mexico (to the east of central USA).

Description:

- one of the most managed rivers in the world, yet still prone to flooding
- an important transport route for goods from the central states to the south
- needs dredging to remain navigable and controlling to stop flooding
- major floods this century have occurred in 1927, 1973 and 1993
- river defences – especially levées (embankments) – prevent most flooding and have encouraged people to build, live and work on the river's flood plain
- over 200 dams have been built on tributaries to hold back water flow
- overflow channels are dug to take some excess flood water
- areas like the Tennessee Valley have been reforested to increase interception and decrease surface run-off
- wing dykes have been constructed to keep a deep river channel open on one side of the river
- sections of the river have been straightened and meanders 'cut off'

The 1993 floods:

- worst ever floods (see table)
- during the spring, rainfall was very heavy
- the ground became saturated, increasing surface run-off
- heavy thunderstorms in late June and early July made the situation worse
- in mid-July several levées broke and flooding became widespread
- bridges were impassable and river traffic stopped
- it took months for the flood water to disperse, ruining buildings and farmland

Deaths	45
Evacuees	74 000
Land flooded	26 000 km^2
Crop losses (1993 crop)	\$6.5 billion
Houses flooded	45 000
Overall estimated damage	**\$10.5**

The Mississippi is a good example of a managed river basin. Yet the most recent floods were the worst ever! Some people argue that attempts to control the river have made things worse … but the control measures do prevent the majority of floods.

EXAM PRACTICE

Question A

a These are the five main processes of the hydrological cycle:
- precipitation
- condensation
- transpiration
- surface run-off
- evaporation.

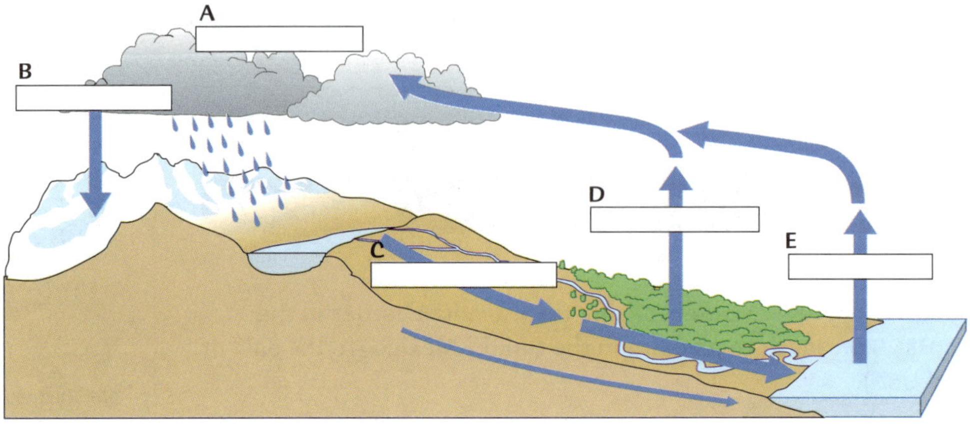

Complete the diagram by putting the labels in the correct boxes.

b Give two reasons why trees play an important part in the hydrological cycle.

c Look at the hydrographs A and B.

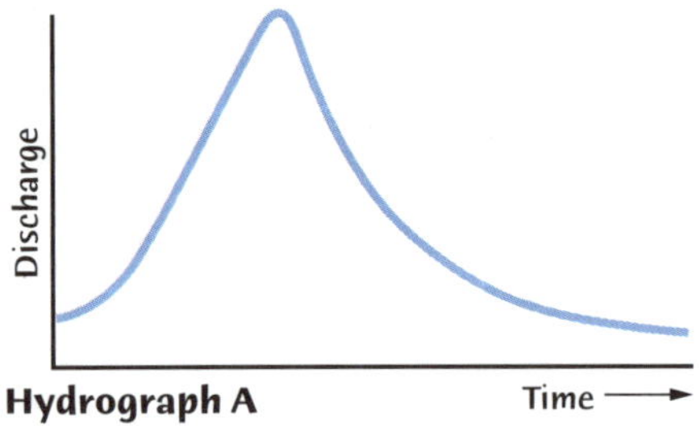

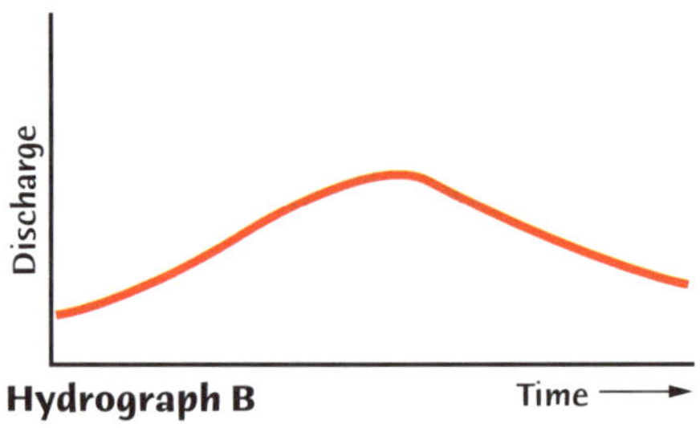

i) Which river is the most likely to flood? Why?

ii) Describe and explain four different factors which could account for the differences between hydrographs A and B.

Question B

Study the OS map extract on p. 92.

a i) In which general direction is the River Parrett flowing?

ii) Name the feature found at 275409.

b Describe in detail the course and features of the river from 303390 to 285450.

c For one of the features you named in (b) draw annotated diagrams to explain its formation.

d What is the main land use in this section of the river valley? Give map evidence to support your answer.

Question C: case study question

For a named river management / water management scheme you have studied:

a Give the name and location.

b Describe the scheme.

c Explain why the scheme was needed.

d Describe the good and bad effects of the scheme.

Revision Unit 3

Ice

The Check up

1 Small ice sheets are called __________.

2 Thousands of years ago much __________ areas of the Earth's surface were covered by ice.

3 The only parts of the UK not covered by ice during the last Ice Age were the __________ and the south west.

4 The effect of ice on the landscape is called __________.

5 Glaciers are moving rivers of ice fed by snow which accumulates in __________ high up in mountain areas.

6 Look at diagram (a). Complete the labels.

7 Plucking occurs when ice freezes and __________ to rock, pulling it away when the ice __________.

8 See diagram (a). A corrie is left when the ice in this diagram melts. Sometimes it is filled by a small lake called a __________.

9 Corries have __________ steep sides or walls and a __________ open side where the glacier moved down the mountainside.

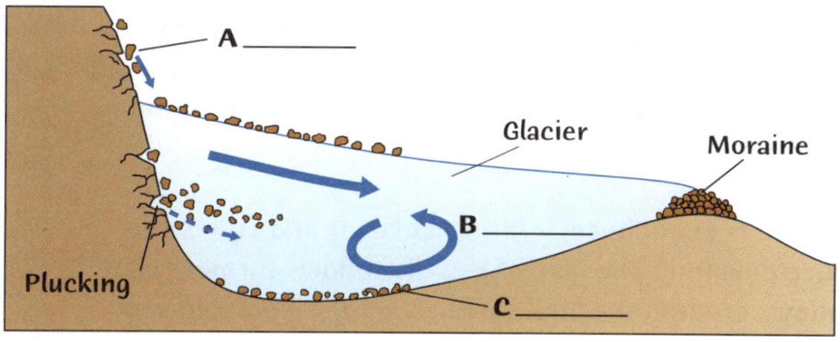

Diagram (a)

10 Where two corries are formed side by side, a steep narrow ridge called an __________ separates them from one another.

11 Where three or more corries are back to back a __________ may form in the middle.

12 U-shaped valleys are typical of glaciated areas. These were once __________ river valleys which have been __________ and __________ by erosion.

13 Glacial valleys are sometimes occupied by __________. These are long and __________ in shape.

14 __________ valleys are small river valleys high up the side of a glacial valley, above the level of the __________.

15 __________ can easily be built across U-shaped valleys to store water behind them, in __________.

16 As glaciers melt they __________ the vast amount of material which they have eroded and transported.

17 See diagram (b). Label each of the moraines.

18 __________ moraines are found at right angles to the flow of the glacier and are left behind when a retreating glacier is stationary for a short time.

19 When glaciers start to melt, __________ streams carry sand, gravel and other material away from the end of the glacier, forming __________ plains.

20 A __________ is a long, narrow hill with one long, gently sloping side and one short steep side.

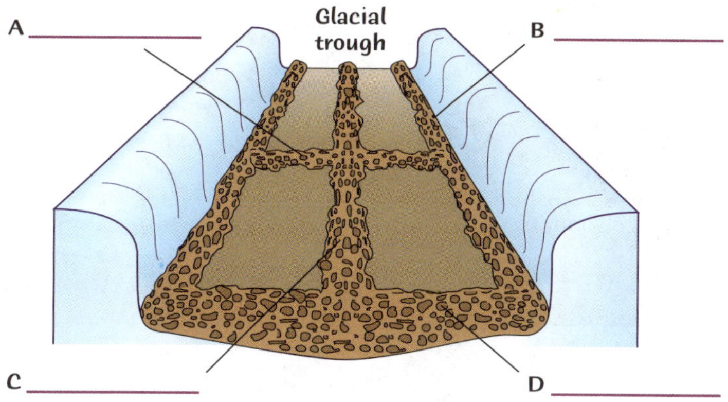

Diagram (b)

Mark yourself (see answers on page 97)

- More than 12? Do the Exam Practice – or go on to another unit.
- Less than 12? Do the Crash Course on the next page.

/20

The Crash Course

Today, ice sheets cover the world's polar regions, with smaller ice caps in areas like Iceland. In high mountain areas snow accumulates, feeding **glaciers** – rivers of ice which occupy highland valleys. Until about 10 000 years ago far greater areas were covered by ice, including all of the British Isles apart from the South and South West. As the ice retreated at the end of this Ice Age it left behind a great variety of landforms – evidence of **glaciation**.

Glaciers

When the climate is cold enough, precipitation falls as snow. This is common in high mountain areas where temperatures are permanently low. Snow accumulates in slight hollows, gradually compacting or squeezing the bottom layers into ice.

If there is enough ice, it forms a glacier and starts to move downhill. The glacier usually follows former river valleys, eroding the land surface and transporting vast amounts of eroded rock. When the ice melts it deposits this material together with any rock debris which falls onto the ice from the surrounding mountain peaks.

A mountain glacier occupying a highland valley

Processes of glacial erosion

The diagram shows the main processes by which glaciers erode.

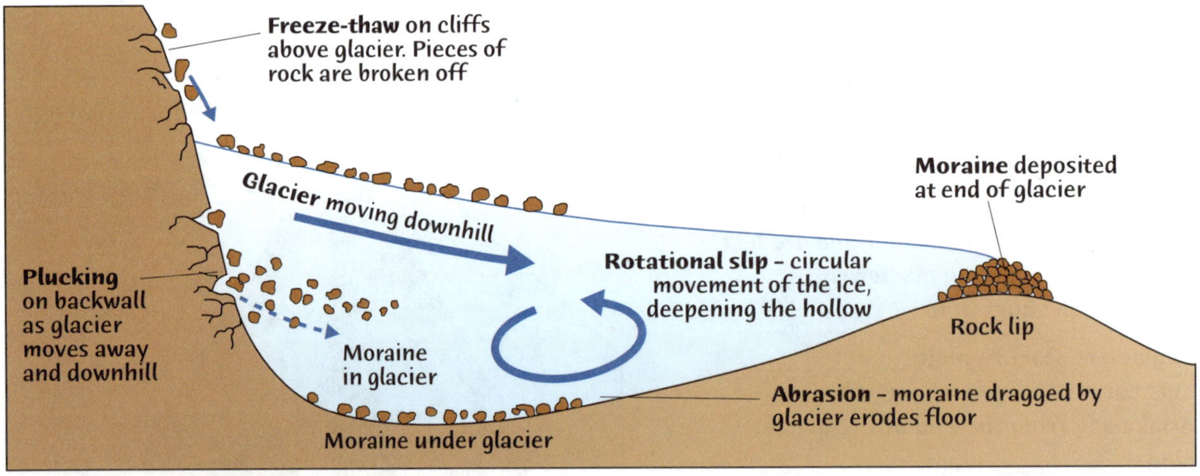

Key facts to learn: processes of glacial erosion

- **Freeze-thaw**: water in cracks in the rock freezes and expands. As it warms up it melts – repeated freezing and thawing causes the rock to shatter
- **Plucking**: ice freezes and sticks to rock, pulling it away when the ice moves
- **Abrasion**: material (pieces of rock) carried by a glacier scrapes away the valley floor and sides
- **Rotational slip**: the circular movement of ice accumulated in a snowfield, deepening the hollow it occupies.

THE CRASH COURSE

Landforms caused by glacial erosion

The diagram on page 22 shows the starting point of a glacier. As the glacier moves down the mountain, it deepens the hollow, forming a feature called a corrie, cirque or cwm. This has three high / steep sides and an open fourth side where the glacier moves out. As the ice melts, a lake (tarn) is often left behind.

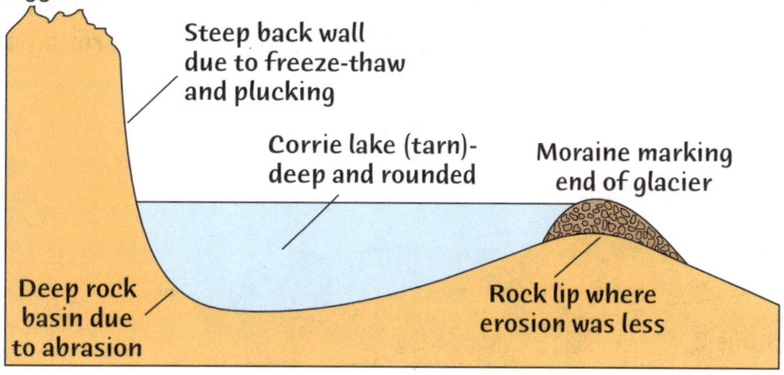

A corrie

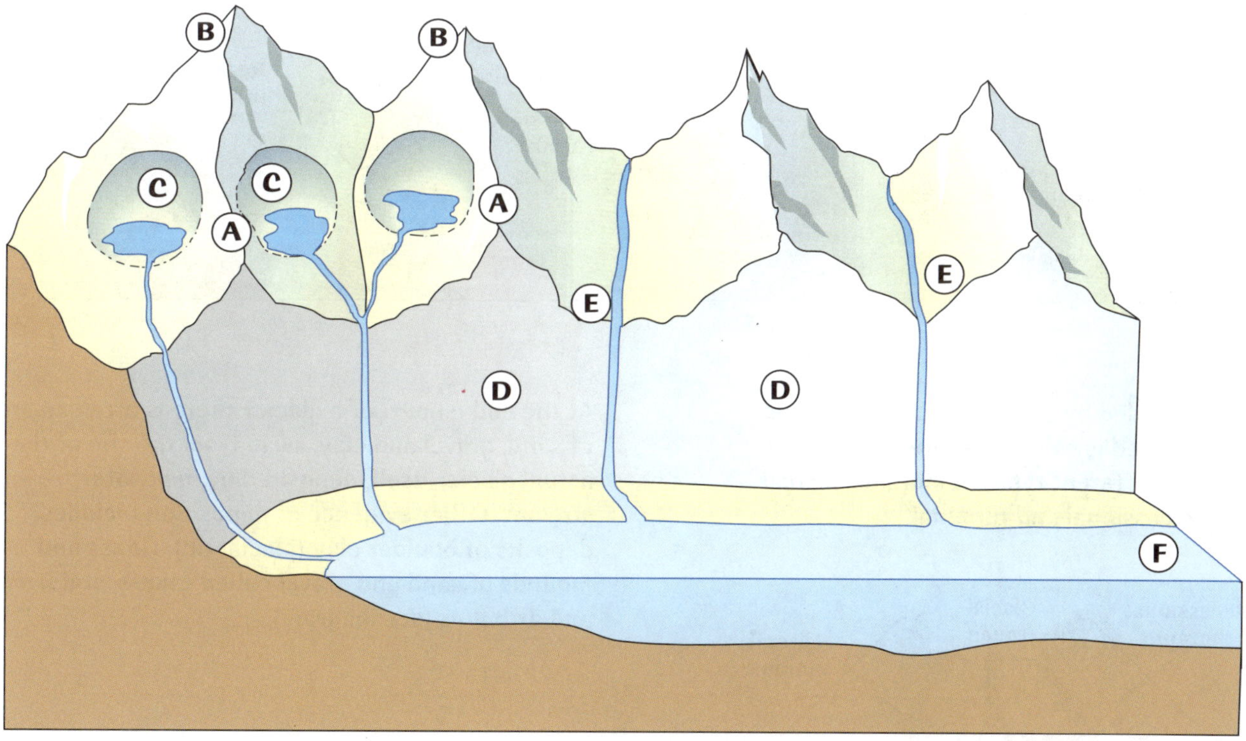

A	**arête**:	steep, narrow, knife-edged ridge separating two corries
B	**pyramidal peak**:	pyramid-shaped mountain peak formed as three or more corries cut back towards each other
C	**corrie**:	rounded hollow in the mountainside where a glacier is formed
D	**U-shaped valley**:	former V-shaped river valley whose sides have been steepened (spurs truncated) and floor deepened by the weight and erosional powers of a glacier
E	**hanging valley**:	V-shaped valley above the level of the ice
F	**ribbon lake**:	long narrow lake occupying a glacial valley; they often form behind glacial deposits at the end of the valley

The Crash Course

Landforms caused by glacial deposition

Glaciers erode and transport vast amounts of material, from large boulders to fine particles. When the temperature rises, glaciers start to melt and retreat back up into highland areas. The eroded material is then deposited, forming a number of distinctive landforms and giving evidence of glaciation.

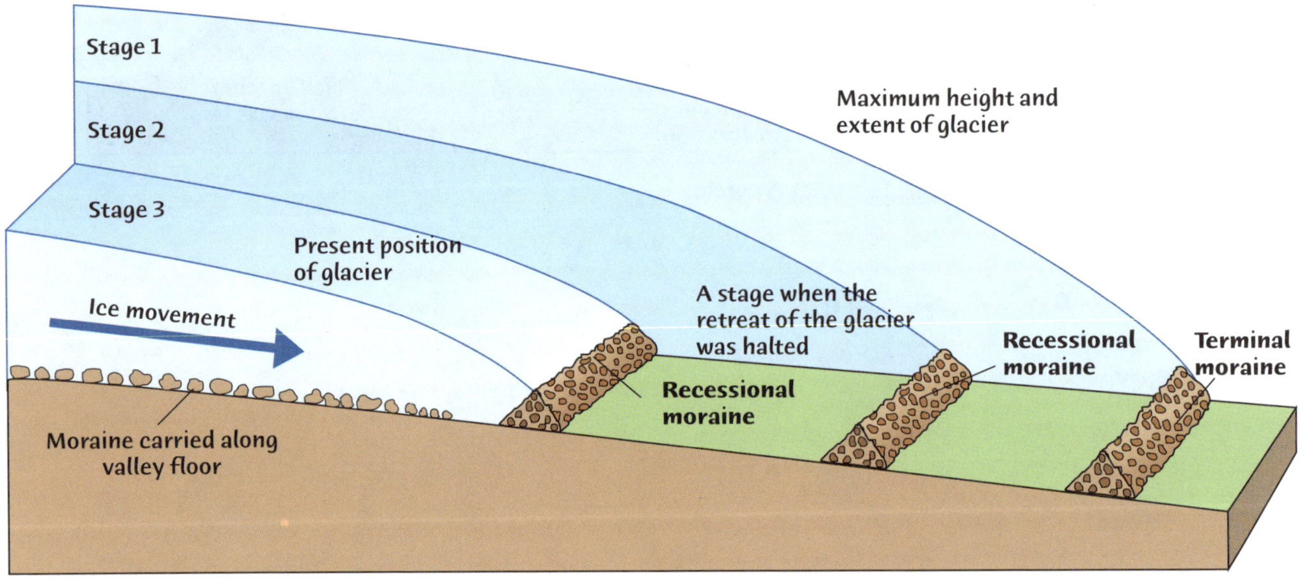

Material eroded by a glacier, transported and then deposited is called **moraine**. There are a number of different types of moraine: medial, lateral, recessional and terminal.

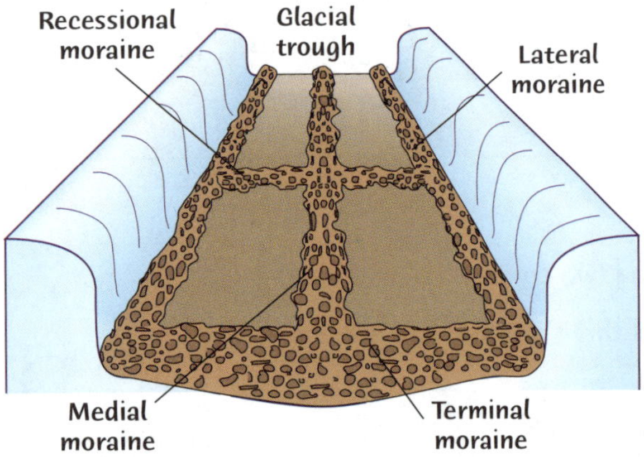

At the end (snout) of a glacier there is often an area of sand, gravel and clay, away from the site of the actual glacier itself, deposited by meltwater streams. Other evidence of deposition includes deposits of boulder clay (glacial till); ridges and mounds of sand and gravel called eskers; and kames and drumlins (see diagram).

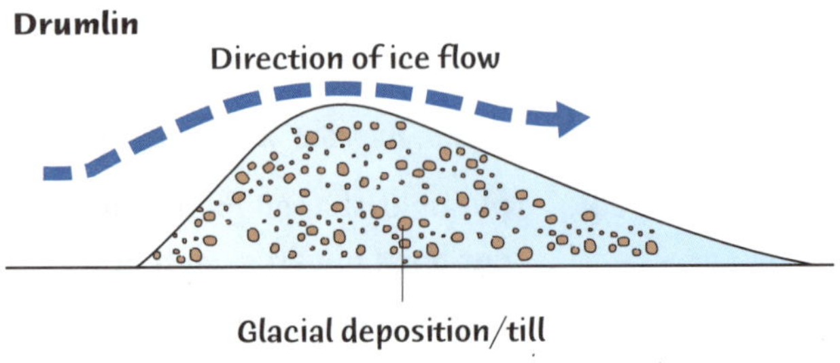

Key facts to learn: types of moraine

- **medial**: found in the centre of a glacier, formed from lateral moraine when two glaciers meet
- **lateral**: found along the edges of a glacier
- **recessional**: found across a glacier, formed as it stops retreating for a while
- **terminal**: deposited at the end of the glacier

Case Studies

Examples

Commonly studied case studies:

UK:
Lake District; Cairngorms (Scotland); Snowdonia (North Wales)

MEDCs:
Denmark; Alps (France, Switzerland, Austria); Rocky Mountains, USA; Alaska, USA

Sample case study: The Lake District, north-west England

Striding Edge and Red Tarn, Helvellyn

Location:

The Lake District is in north-west England, west of the Pennines and close to the Scottish border.

Description:

- the Cumbrian mountains form the centre of the Lake District
- during the Ice Age glaciers occupied the mountains and valleys
- today U-shaped valleys with ribbon lakes radiate out from the centre
- the three highest peaks are Helvellyn, Scafell Pike and Skiddaw

Examples of glacial features are:

- corries and tarns e.g. Blea Tarn and Red Tarn
- arêtes e.g. Striding Edge (above Red Tarn); High Street and Long Stile (above Blea Tarn)
- ribbon lakes e.g. Windermere, Coniston and Derwent Water
- terminal moraine e.g. damming the end of the valley occupied by Lake Windermere
- drumlins e.g. Eden Valley (to the east)

In studying a glacial area it is also usual to look at the impact of human activity on the landscape – and vice versa. In particular, since these areas are usually very attractive, large numbers of tourists visit them. This can cause conflict between visitors, landowners, local people and any industrial activity taking place there.

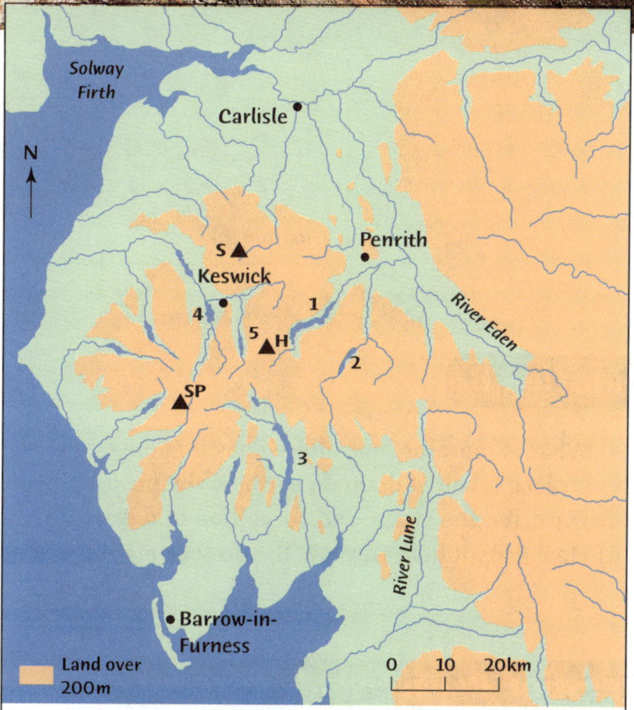

Key
1 Ullswater
2 Hawes Water
3 Windermere
4 Derwent Water
5 Thirlmere
H Helvellyn
SP Scafell Pike
S Skiddaw

Exam Practice

Question A

a Describe two of the main processes of glacial erosion – plucking and abrasion. You may use diagrams to help.

b Study the three diagrams showing freeze-thaw weathering (a). Describe how this process causes rocks to shatter.

c Look at diagram (b). Explain how the three main features labelled A,B and C are formed.

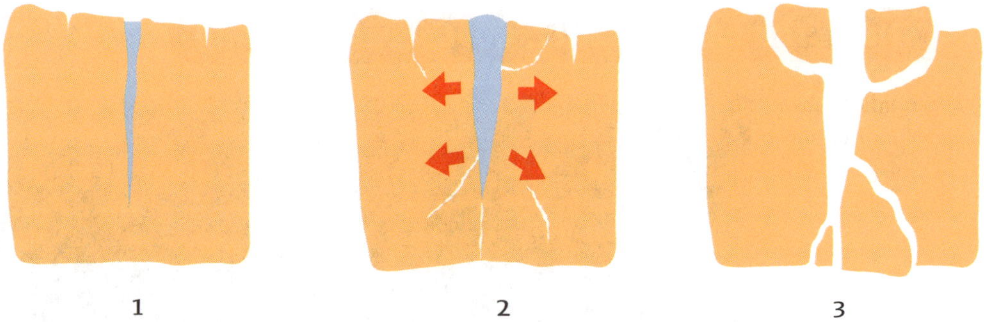

Diagram (a)

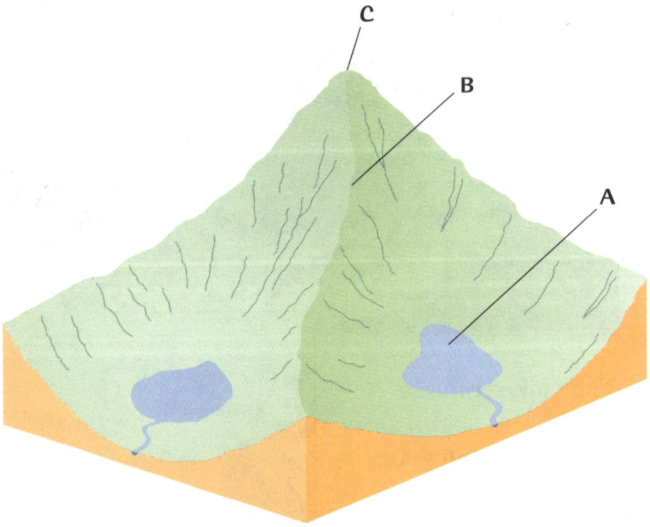

Diagram (b)

Question B

a i) What is moraine and where does it come from?
ii) Draw a labelled diagram to show the position of the four different types of moraines.

b i) Describe the shape and formation of a drumlin.
ii) How does it help to show the direction of ice movement?

Question C: case study question

a For a glaciated area you have studied, name the area and draw a simple sketch map to show its location and main features.

b List examples of either glacial deposition or glacial erosion in your named area.

c What conflicts are there between different groups of people or users?

Revision Unit 4

Coasts

The Check up

1 ___________ is the distance across the sea that winds blow to produce waves.

2 ___________ waves are high and steep and cause erosion.

3 ___________ waves are lower and usually transport and deposit material.

4 The ___________ is the part of the wave which pushes up a beach; the ___________ takes water back down the beach.

5 Coastlines are shaped by three main processes – erosion, ___________ and ___________.

6 Material can be moved along a coastal area by a process called ___________ ___________.

7 For the process in question 6 to occur, waves must approach the land at an ___________.

8 ___________ is the process by which rocks and pebbles are worn down by grinding against each other.

9 Corrosion is a process of erosion where sea water erodes land by ___________ chemicals in the rock.

10 Headlands may form when ___________ rock is found on the coast. This is because it is more ___________ to erosion.

11 Softer rocks are less resistant and are eroded ___________ where they meet the sea.

12 Cliffs are eroded and gradually move inland as waves erode a ___________ at the base, eventually causing overhanging rock to ___________.

13 As cliffs are worn back they leave behind a rocky feature called a ___________ which is visible at low tide.

14 Look at diagram (a). Add the correct labels in the spaces provided.

15 When an arch collapses it leaves a feature called a ___________.

16 If a spit grows and completely blocks the mouth of a river it is called a ___________.

17 ___________ are often built at right angles to the coast to stop beach material from being ___________ down the coast.

18 Look at diagram (b) then complete the following: Building the concrete wall at ___________ will help keep area ___________ supplied with beach material, but will stop it from reaching ___________ as it will gradually build up to the left of ___________.

19 Beaches are important because they absorb ___________ from waves as they hit the land.

20 Wooden, slatted structures called ___________ are sometimes built to absorb the energy of waves as they break.

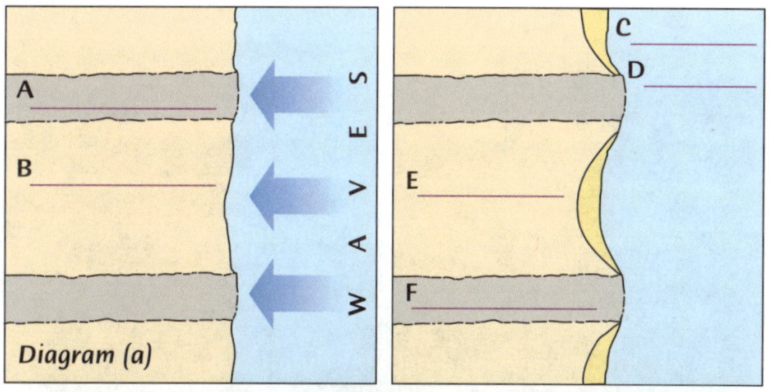

Diagram (a)

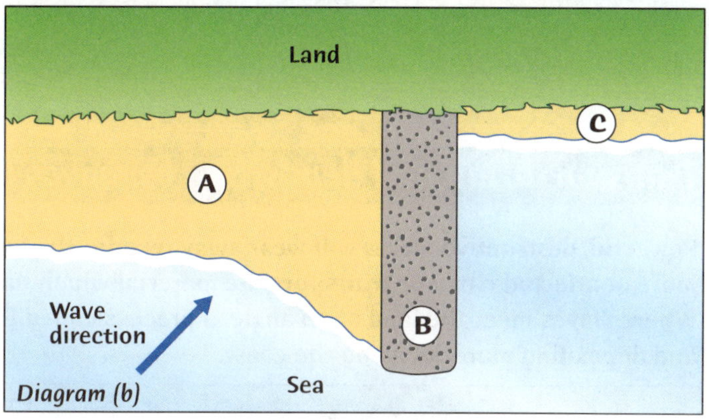

Diagram (b)

Mark yourself *(see answers on page 97)*

- More than 12? Do the Exam Practice – or go on to another unit.
- Less than 12? Do the Crash Course on the next page.

/20

THE CRASH COURSE

The work of the sea: processes

Coastal areas are constantly changing because of the action of the sea. As waves break on the land they may erode, transport or deposit material, shaping the coastline as they do.

Wave action

The height, speed and strength of waves depends on:

- the strength of the winds
- the amount of water (distance) the wind blows across before reaching land (the **fetch**)
- the amount of friction between the water, air and land surface.

As waves reach land the top part (the **swash**) topples (breaks) over and washes up the beach. The water then runs back down the beach (**backwash**) to the sea. There are two main types of waves:

Storm waves (destructive) eroding the coast

Constructive waves

- 6–8 per minute
- swash stronger than backwash
- low height and gradient
- deposit rather than erode

Destructive waves

- 10–12 per minute
- backwash stronger than swash
- high, steep waves
- erode the coastline

KEY FACTS TO LEARN: PROCESSES OF WAVE EROSION

There are four main processes of erosion carried out by waves:

- **Abrasion / corrasion**: waves throw sand, pebbles and rock particles against the land surface, rubbing and grinding it down
- **Attrition**: rocks and pebbles bump and grind against each other, so that they are rounded down and worn away
- **Hydraulic action**: as waves pound against the land air is forced into and trapped in spaces or cracks in the rock. The pressure of the compressed air can cause the rock to break up
- **Corrosion**: acidic saltwater may dissolve minerals or chemicals in certain rocks

Transportation and deposition by waves

Powerful, destructive waves will wear away (**erode**) the land. However, for much of the time waves are more gentle and carry or **transport** the material which has been eroded, depositing it along the coastline. Where waves meet the land at an angle, a process called **longshore drift** is responsible for transporting and depositing material along the coast.

The Crash Course

Landforms caused by coastal erosion

Headlands and bays

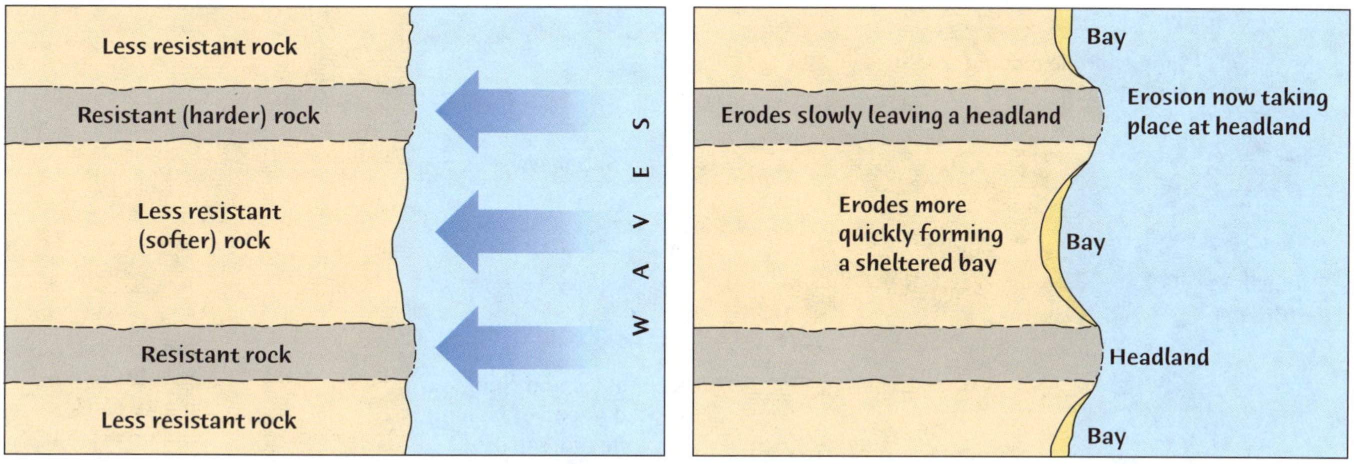

The two diagrams show that for headlands and bays to form in coastal areas there must be rocks of different hardnesses. Hard rock is more resistant and will erode more slowly than softer, less resistant rock.

Cliffs and wave-cut platforms

Where hard rock meets the sea, usually as a headland, cliffs are likely to form. Erosion takes place where the rock is in contact with the sea, at the base of the cliff. Over time the cliff will be eroded and move further inland, leaving behind a rocky surface called a wave-cut platform. Erosion takes place constantly, but in stormy weather, with high waves, erosion can be rapid and cliffs may collapse suddenly.

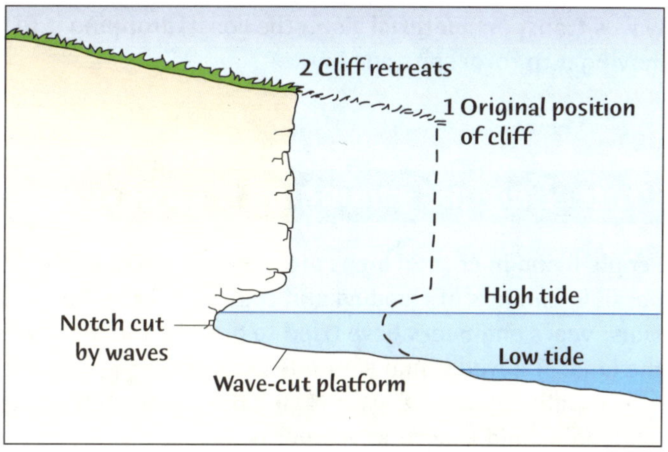

Caves, arches, stacks and stumps

The diagram below shows the sequence of erosion of a headland, producing a number of very distinctive coastal landforms.

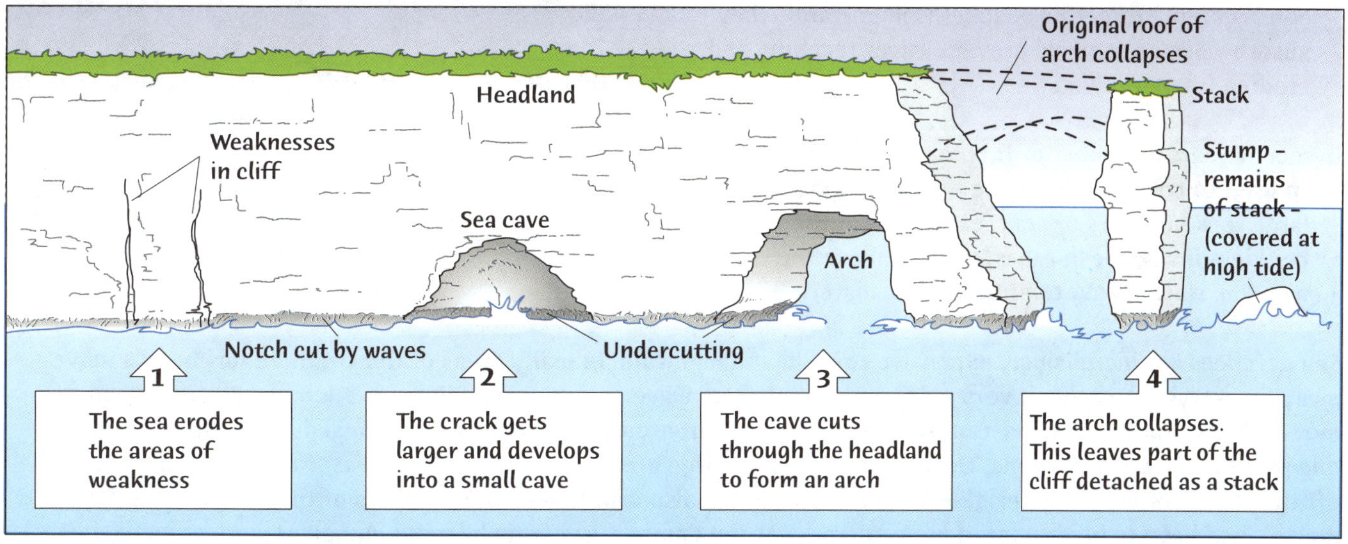

The Crash Course

Landforms caused by coastal deposition

Beaches, spits and bars

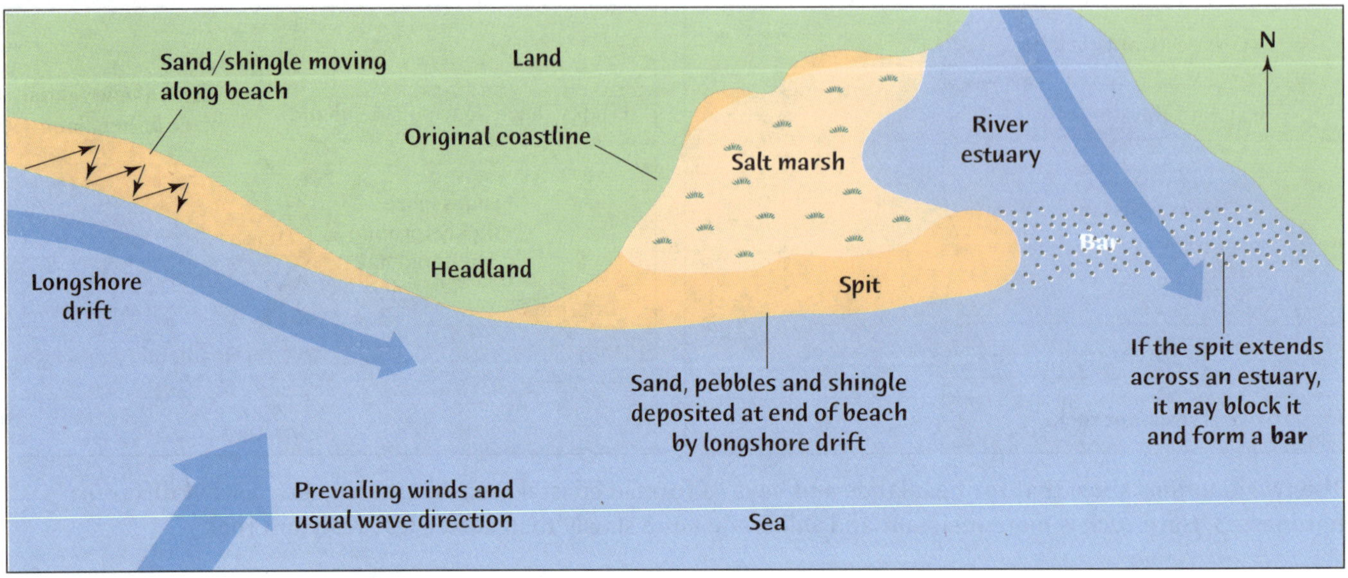

Waves transport material along the coast, dropping it to form beaches and, under the right conditions, moving it to form spits and bars.

Protecting and managing the coast

People living in coastal areas are only too aware of the possible dangers of flooding and coastal erosion. For many years engineers have tried to protect and manage the land. The traditional methods used are:

- sea walls – concrete walls facing the sea to stop flooding and absorb wave energy;
- groynes – wooden walls built at right angles to beaches to prevent beach material from being taken down the coast by longshore drift. Sometimes much larger concrete breakwaters (like large groynes) are constructed. Protecting beaches is important – they absorb wave energy and prevent waves reaching and eroding further inland.

Groynes preventing long-shore drift at Southwold, Suffolk

In recent years a number of further methods have been used to try to absorb wave energy and stop coastal erosion:

- large rock boulders or concrete blocks at the top of beaches;
- boulders inside mesh cages, usually above the sea wall;
- wooden, slatted **revetments** (to trap material).

Sea defences are increasingly expensive to build and maintain. In many areas of the UK there has been a move away from trying to protect very vulnerable areas from wave action. People living in such places are having to move away to safer locations. Conflict between different groups of people can be caused by action taken along one part of coast. For example, the construction of a large breakwater affects the coast further down by cutting off the supply of beach material. Areas of coastline are also visited by large numbers of tourists and holiday makers and have to be protected not just from natural damage, but from harmful human activity as well.

CASE STUDIES

EXAMPLES

Commonly used case studies:
UK:
local study; NE coast (England) – Scarborough, Holderness and Mappleton;
south coast (England) – Swanage, Dorset, Portland, Purbeck, Isle of Wight, Devon

Sample case study: Mappleton, Holderness, UK

Location:

A 40km stretch of coastline from Hornsea to Spurn Head, along the North Sea coast of north-east England.

Description:

- the cliffs forming this part of the coast are 20–30 metres high
- they are made from soft, unconsolidated clays and sands
- erosion is rapid in soft rocks, as is the likelihood of slipping and slumping
- waves usually approach the coast at an angle from the north east
- longshore drift moves vast quantities of material southwards
- 1.5 million tonnes of mud is washed out to sea every year
- the mud and sand is eroded from the cliffs
- sea level is slowly rising along this coast

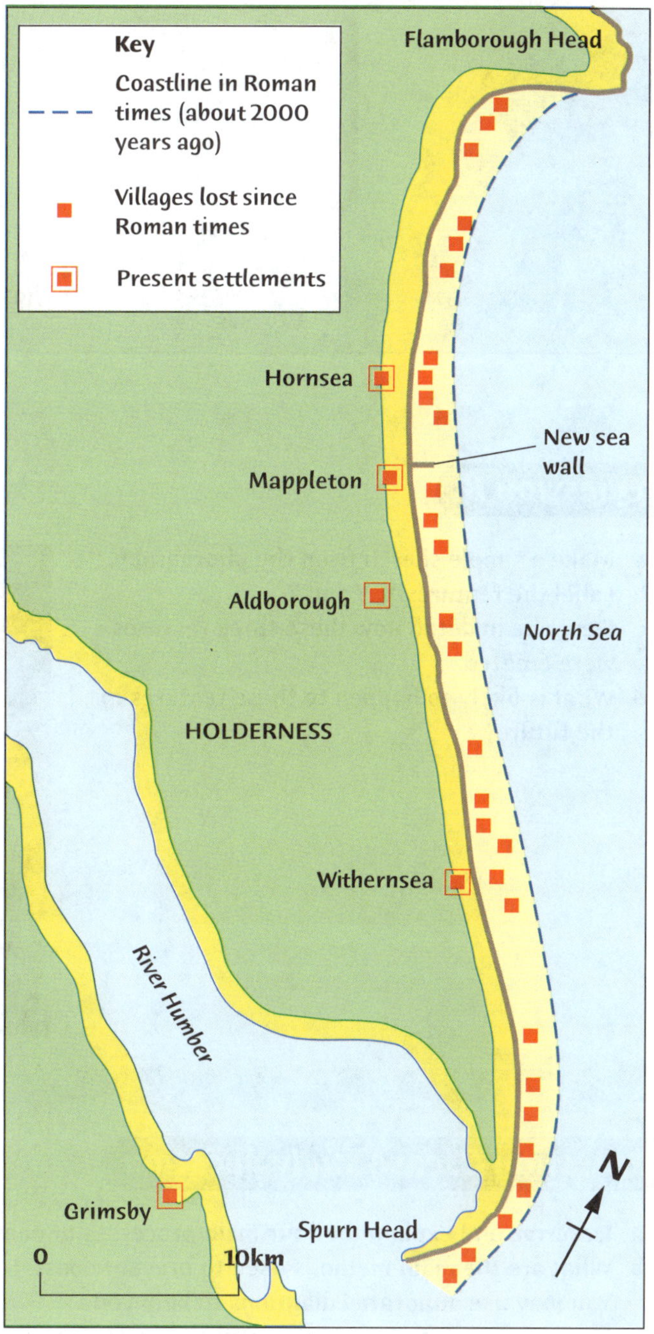

Protecting Mappleton:

- in 1991 a large sea wall was built at Mappleton, at right angles to the land
- the wall cost £2 million to construct
- it was designed to stop the beach being eroded and transported south by the waves, helping to protect the cliffs from erosion
- the wall is successful – for Mappleton
- the wall has cut off the supply of beach material to the south
- with no new beach material the cliffs are unprotected and erosion south of the sea wall has increased sixfold
- houses and agricultural land are being 'lost' to the sea and people are having to move away

This is a good example of human action along one stretch of coastline having an impact on coastal processes elsewhere. There are many examples of this along the UK's coastline.

EXAM PRACTICE

Question A

a Draw an annotated diagram to show the process of longshore drift.

b Look at the diagram. Explain:

i) the pattern formed by the beach material;

ii) why the groynes are built.

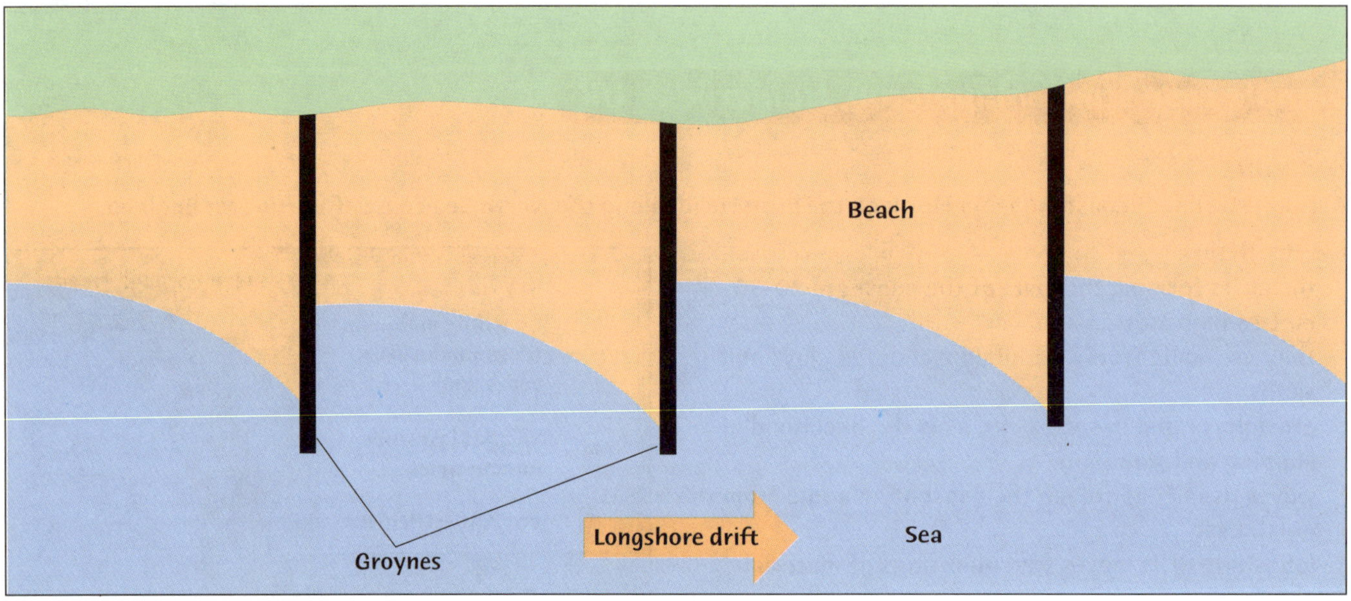

Question B

a Make a simple sketch from the photograph.

b Label the features X, Y and Z.

c Describe in detail how these three features were created.

d What is likely to happen to these features in the future?

Question C: case study question

a Describe and explain the four main processes of coastal erosion.

b What are the main methods used to prevent coastal erosion? You may use annotated diagrams to help you.

c Name an area you have studied where coastal erosion is a problem. What is being done to try to prevent or reduce erosion there?

Revision Unit 5

Weather and climate

The Check up

1 Weather is the daily state of the __________.

2 Climate is the __________ weather over __________ years.

3 The two main influences on climate are __________ (distance from the Equator) and __________ (height above sea level).

4 Climate graphs show both __________ (line graph) and __________ (bar graph).

5 The range is the difference between the __________ and __________ temperatures.

6 Much of the UK's weather comes from the __________ because of direction of the __________ winds.

7 Places nearer the __________ usually have a smaller range of temperature than those further __________.

8 The east coast of the UK is generally __________ and __________ than the west in the summer.

9 The further away from the Equator you go, the colder it gets. This is because
____.

10 The main elements (types) of weather measured are temperature, __________, humidity, __________ (speed and direction), __________, sunshine and cloud.

11 Wind speed is measured by an instrument called an __________.

12 Describe the weather at the station (diagram (a)):

Temperature: ______________________________

Wind speed: ______________________________

Wind direction: ______________________________

Weather conditions (precipitation): ______________________________

Cloud cover: ______________________________

Diagram (a)

13 When water vapour rises, it cools down and __________, causing clouds and rain.

14 There are three main types of precipitation – convectional, __________ and __________.

15 If a continental air mass moves over the UK in summer it would bring __________ and __________ weather.

16 Look at diagram (b) showing isobars:

i) Label an area of high pressure (H) and an area of low pressure (L).

ii) Draw in an arrow to show which direction the wind would be blowing from.

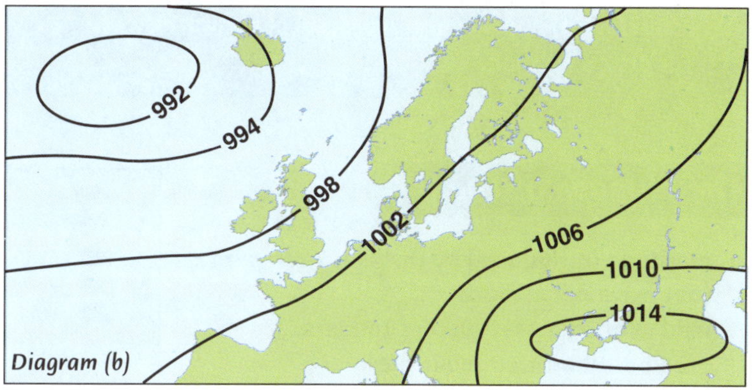

Diagram (b)

17 Low pressure systems are called cyclones or __________.

18 In the UK an anticyclone brings __________ weather in summer and __________ weather in winter.

19 When a warm air mass meets a cold air mass the warm air __________.

20 Tropical storms can develop into __________, __________ or __________.

Mark yourself *(see answers on page 98)*

- More than 12? Do the Exam Practice – or go on to another unit.
- Less than 12? Do the Crash Course on the next page.

/20

THE CRASH COURSE

World climate

Influences on climate

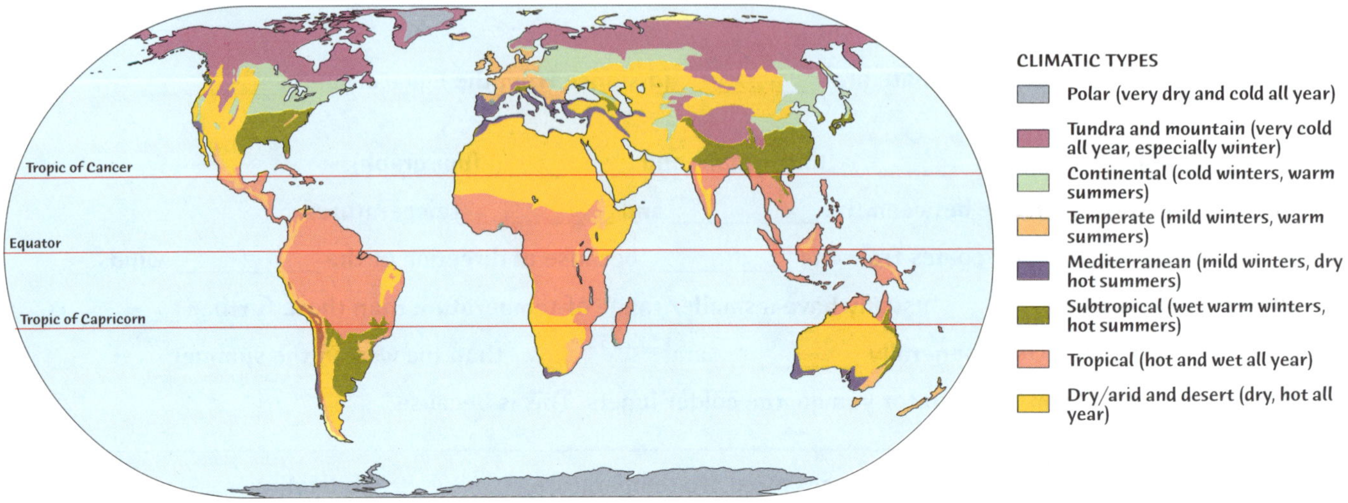

Major world climate types

KEY FACTS TO LEARN: FACTORS AFFECTING CLIMATE

- **Latitude**: temperatures decrease away from the Equator. The curved surface of the Earth means that the polar regions get less heat as they are further from the sun and receive heat at an angle.
- **Altitude**: the higher up you are the colder it gets as the air is less dense and can't hold heat as easily. Temperatures fall by 6.5°C for every 1000 metres above sea level.
- **Winds**: strong winds can lower temperatures (wind chill factor) and move air masses (cold or warm, wet or dry) across land surfaces.
- **Distance from the sea**: land and sea heat up / cool down at different rates. Places near the sea are warmer in winter and cooler in summer (small temperature range), places inland are the opposite (large temperature range).
- **Aspect**: slopes facing the sun are warmer e.g. south-facing (northern hemisphere) and north facing (southern hemisphere).

The climate of the UK

- maritime (influenced by the sea) / temperate (not extreme) climate
- mild winters, cool / warm summers, precipitation throughout the year
- changeable weather
- prevailing westerly winds
- dominated by depressions from the west
- precipitation is highest in highland areas of the north and west
- temperatures in winter = warmest in SW, coldest in NE; in summer = warmest in south, coolest in north

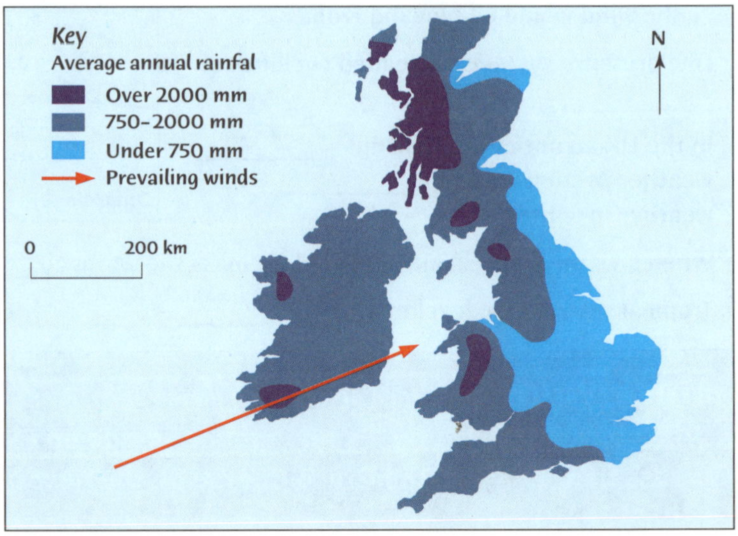

The Crash Course

Air masses and weather systems

Air can be hot or cold, wet or dry. As it moves from one place to another it takes those characteristics with it. Air moving from the poles will be cold; from the Equator it will be warm. If it travels over water it will be moist; over land it may be dry. Winds are responsible for moving these air masses over the Earth's surface, creating and changing weather conditions.

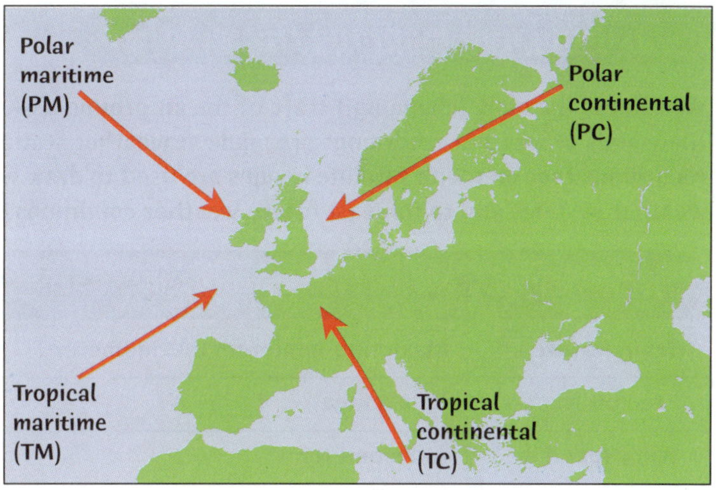

Cyclones or depressions (areas of low pressure)

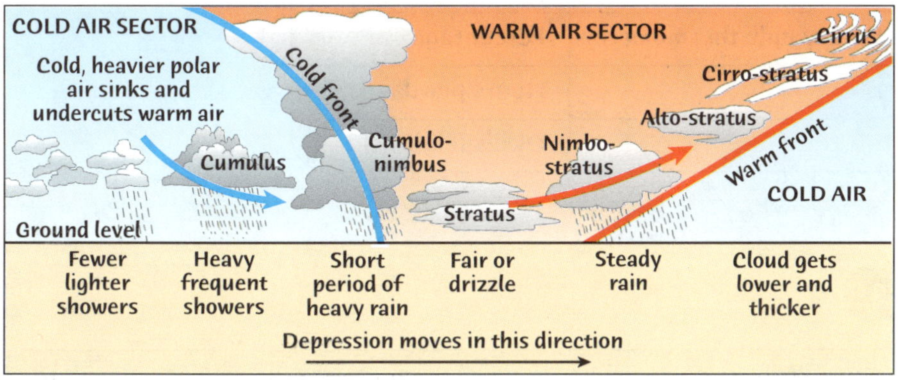

Anticyclones (areas of high pressure)

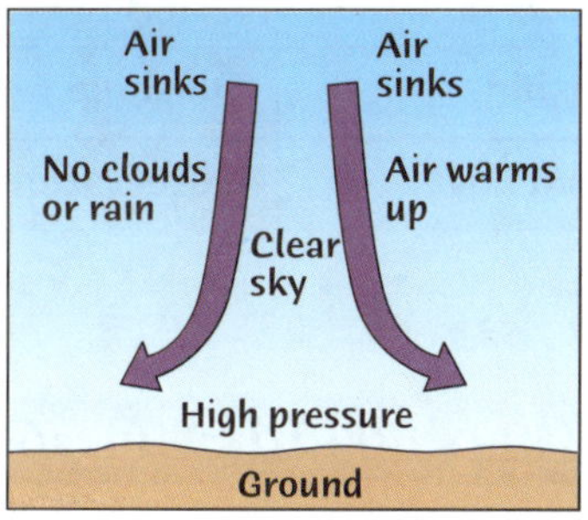

Types of rainfall

The diagram of a depression shows a type of rainfall called frontal rainfall which forms where cold and warm air fronts meet. The other two types are:

Relief or orographic rainfall

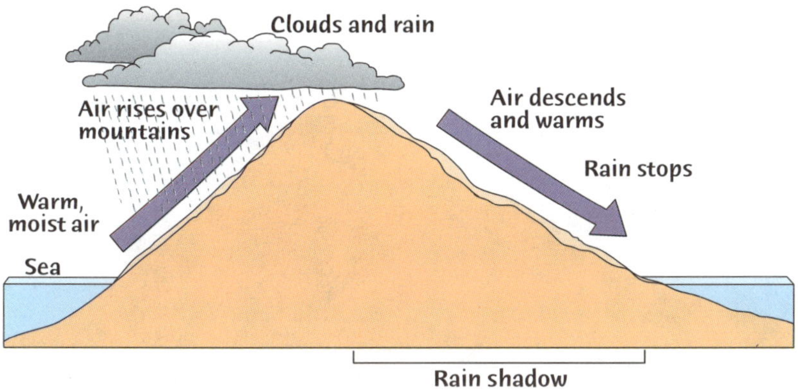

This type of rainfall is typical along the west coast of the UK. The east lies in an area called a **rain shadow**.

Convectional rainfall

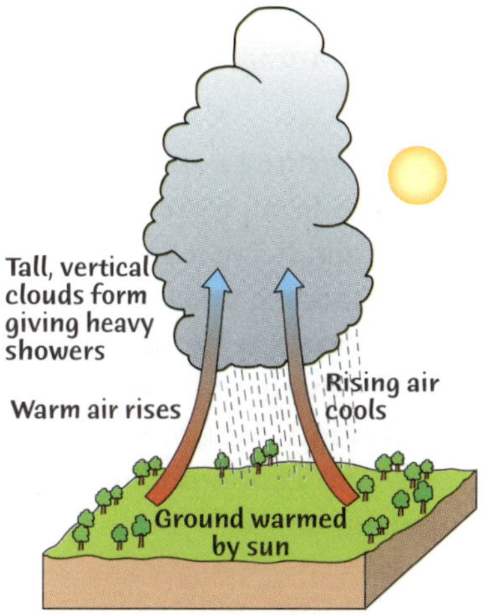

This type of rainfall is typical of thunderstorms and tropical areas.

Tip: Weather is the state of the atmosphere at a particular time. Climate is the average weather over a long period of time (20-30 years).

THE CRASH COURSE

Measuring and mapping the weather

Weather is the daily (changing!) state of the air around us - the atmosphere. Every day thousands of weather recordings are made at weather stations across the world. These recordings together with satellite images are used to draw weather maps, identify weather systems and to forecast future weather conditions.

Type of weather	Weather instrument used	Measured in (units)
Temperature	Maximum-minimum thermometer	°Celsius
Precipitation	Rain gauge	mms
Wind speed	Anemometer	kph/knots
Wind direction	Wind vane	Compass direction
Air pressure	Barometer	Millibars
Humidity	Hygrometer / wet & dry bulb thermometer	Percentage
Sunshine	Sunshine recorder	Hours per day
Cloud cover	Observation	Oktas (eighths)

Weather hazards: hurricanes

Whilst weather can sometimes be a nuisance, it can also bring dangerous and hazardous conditions. Some of the most dangerous weather hazards are hurricanes - also called typhoons or tropical cyclones. The map shows where in the world hurricanes are found.

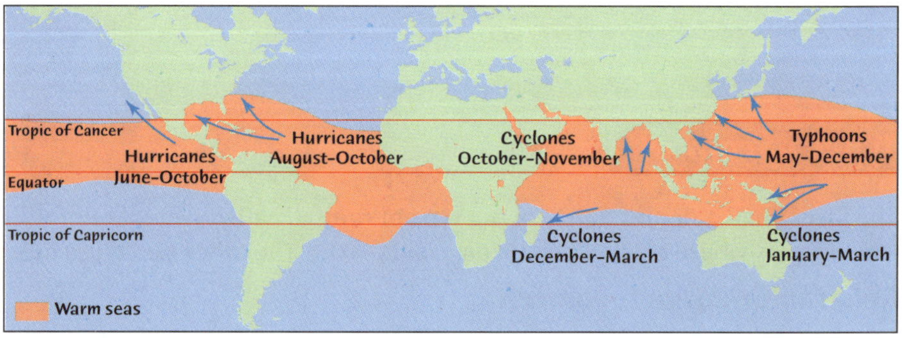

How hurricanes are formed

Hurricanes bring torrential rain and high winds (over 120 km / h) causing flooding and widespread damage and destruction. They can usually be monitored by satellite and warnings given, but few effective precautions can be taken to prevent damage.

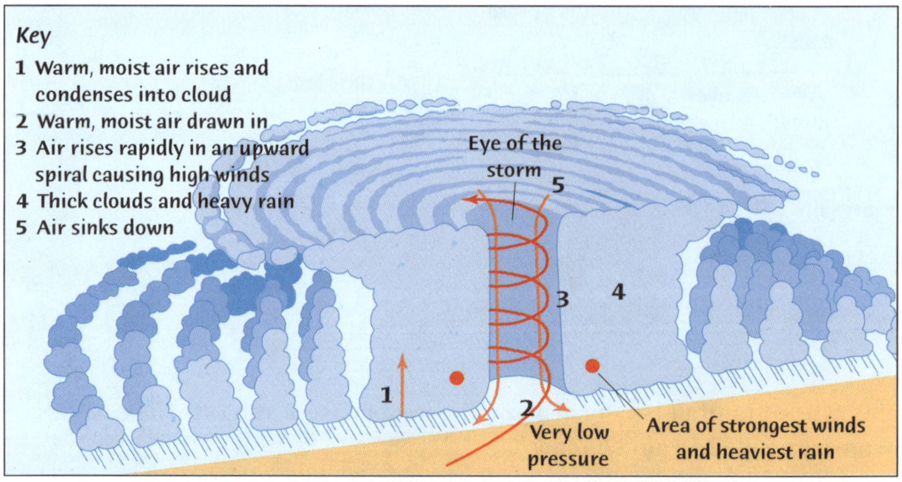

Tip: Other weather hazards - flooding, drought, fires, tornadoes, landslides, mudslides

Case Studies

Examples

Commonly used case studies:
UK climate / depressions and other world climate types e.g. Tropical (Amazon, Brazil); Savanna (Kenya) etc.
Hurricanes: Gilbert, USA (1988); Andrew, USA (1992); Mitch, Central America (1998)
Tornadoes: Texas (1998); Oklahoma (1999), (USA)
Flooding: Lynmouth, UK (1952); Bangladesh; Mississippi, USA (1993)
Drought: UK (1975-76, 1995-96); The Sahel region, Africa (1980's onwards)
Fires: Australia (1993), Indonesia (1998)

Sample case study: Hurricane Andrew

Location

The Caribbean, Florida and Louisiana, USA in August 1992

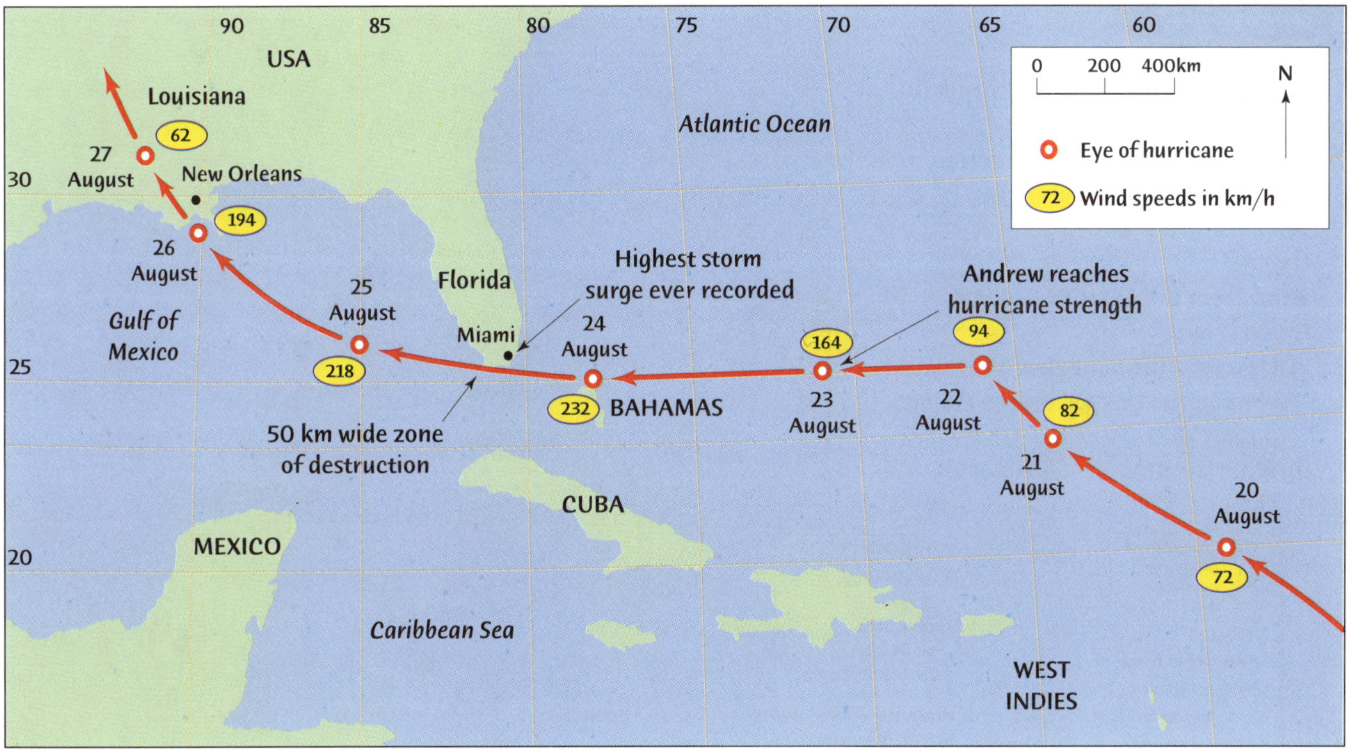

Diary of a hurricane

- 16 August: Andrew starts as a tropical wave off the coast of West Africa
- 17 August: develops into a tropical storm, moving westwards towards the USA
- 21 August: crosses the Tropic of Cancer
- 24 August: started to slow down, then suddenly changed direction to the west
- 22 August: winds reach 230kph (Category 4 hurricane – out of 5). Florida hit as tide rises an extra three metres
- 24 August: highest storm waves recorded off Florida
- 26 August: changes direction, moving north to hit Louisiana
- 28 August: hurricane stops after hitting mainland USA: 26 left dead, 250 000 homeless, $30 billion damage

Although the actual death toll was quite small, Andrew was the most destructive natural hazard in the history of the United States. Hurricane warnings allowed many to board up their homes and leave – but the destruction couldn't be stopped.

EXAM PRACTICE

Question A

Look at the climate graph – diagram (a).

a What is the range of temperature?
b What is the total precipitation?
c Describe the pattern of temperature and climate shown by this graph.
d What climatic type is shown here? Give reasons for your answer.

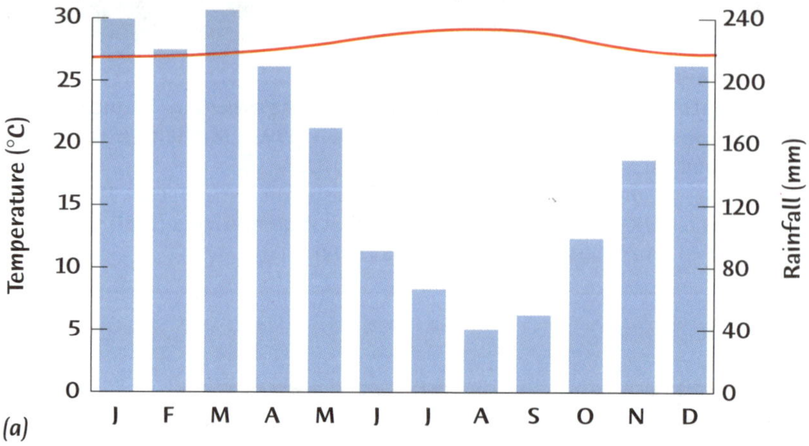

Diagram (a)

Question B

Look at the weather map – diagram (b).

a What is the feature at A?
b Use the following information to complete weather station B:
Temperature 9°C
Drizzle
Wind from the SE at 13–17 knots
Cloud 4 oktas
c i) Describe the sequence of weather changes as this weather system passes over point C.
ii) What weather system does this show?

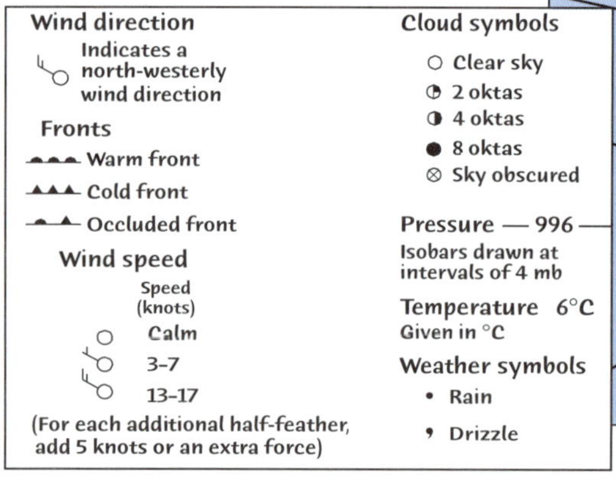

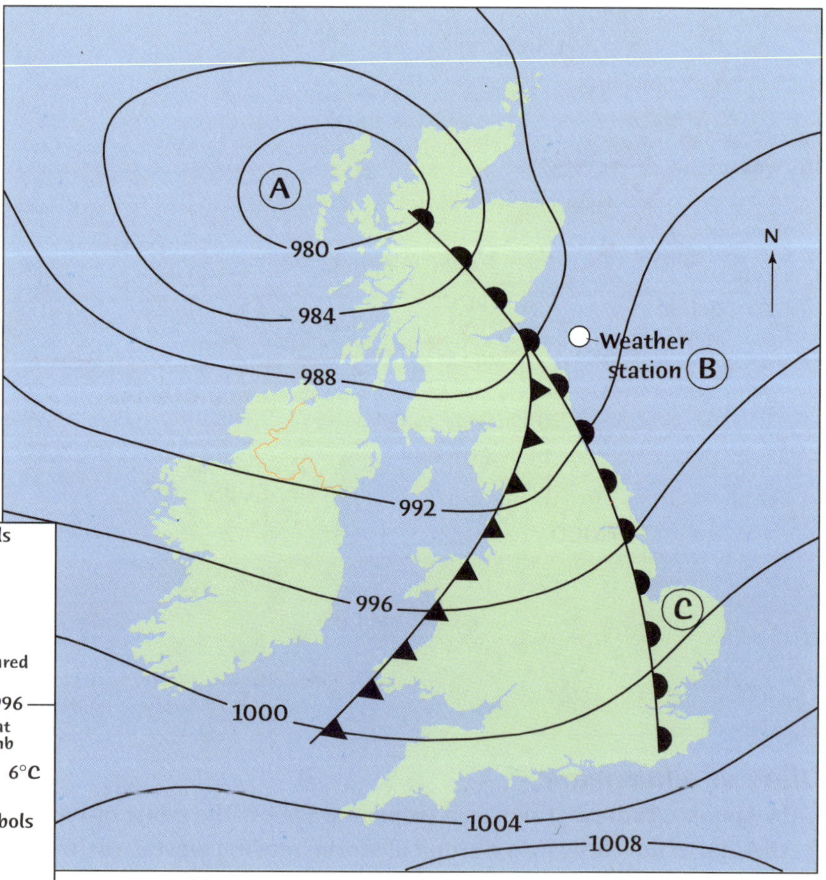

Diagram (b)

Question C: case study question

For a weather hazard / event you have studied:

a describe the causes of the hazard;
b where did it happen?
c what damage was done?
d how was the damage caused?
e what precautions could have been taken to avoid damage and injury?

Revision Unit 6

Ecosystems

The Check up

1 Ecosystems are made of ___________ and ___________ things.

2 They can be broken down into inputs, ___________ and ___________.

3 The main input into an ecosystem is ___________ from the ___________.

4 Plants are at the bottom of the food chain – they are known as primary ___________.

5 The amount of living material in an ecosystem is called the ___________.

6 ___________ are vital to the ecosystem because they break down waste and dead matter and return the ___________ to the soil.

7 Look at diagram (a) and fill in the missing words at A, B and C.

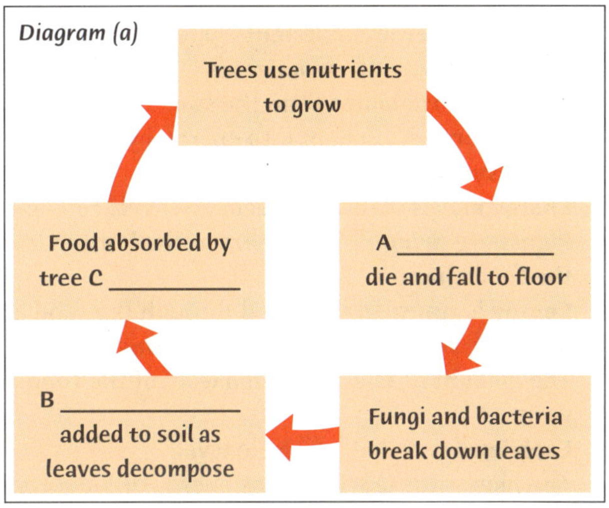

8 The world's major ecosystems are called ___________. The different vegetation found in each is usually the result of different ___________ conditions.

9 Savanna is a major biome found in a zone between hot desert regions and ___________ regions.

10 The typical vegetation of savanna ecosystems is scattered trees, ___________ and ___________.

11 Some savanna vegetation adapts to drought conditions by having ___________ roots and thick, ___________ leaves.

12 Rainforests have the greatest range of ___________ and ___________ in the world.

13 Heavy ___________ rainfall falls most afternoons in rainforests.

14 Large trees which stand out above the canopy of the rainforest are called ___________.

15 Beneath the canopy – the richest area of the rainforest – is the ___________ canopy. Here ___________ tree trunks are covered in ___________.

16 The forest floor is dark, ___________ and ___________.

17 Look at diagram (b) and fill in the missing words at A, B, C and D.

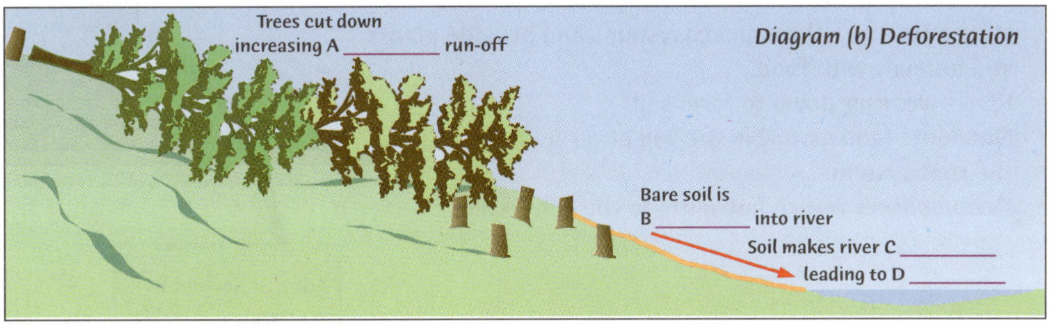

18 Without leaf litter and other decaying plant matter, the soil becomes ___________.

19 Deforestation is having an impact across the world – as trees are destroyed levels of ___________ increase, contributing to global ___________.

20 Rainforest plants are used to provide ___________ and may provide new cures in the future.

Mark yourself *(see answers on page 98)*

- More than 12? Do the Exam Practice – or go on to another unit.
- Less than 12? Do the Crash Course on the next page.

/20

THE CRASH COURSE

Ecosystems

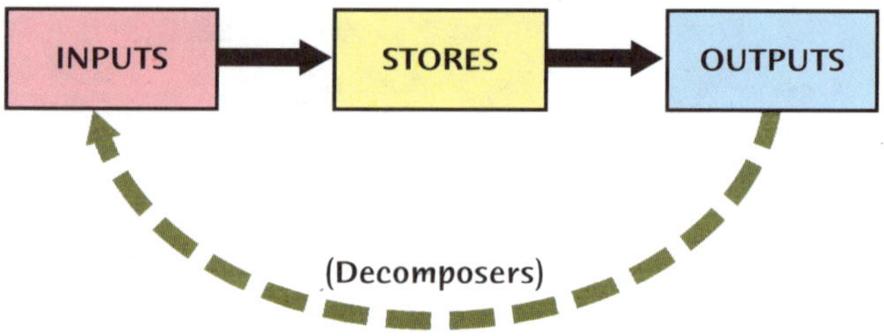

Ecosystems are made up from **biotic** or **living things** e.g. plants, insects, animals, and **abiotic** or **non-living things** (climate, rocks, soil). The different parts work together and are interdependent. Ecosystems can be broken down into **inputs**, **stores** and **outputs**.

Energy flows

- **Energy**, **water** and **nutrients** may move or flow between stores, often in **cycles**.
- All energy originates from the sun.
- Plants convert sunlight into energy by the process of **photosynthesis**.
- Energy moves through an ecosystem via food chains.
- Plants are **primary producers** at the bottom level of the food chain.
- Energy is transferred up and through the food chain via **primary**, **secondary** and **tertiary consumers**.
- The amount of energy at each level of the food chain varies.
- Each level is called a **trophic** level.
- The higher the level, the less energy or biomass is present.
- Decomposers are vital parts of an ecosystem. These are insects, fungi and bacteria which break down waste or dead matter, returning it to the ecosystem via the soil.

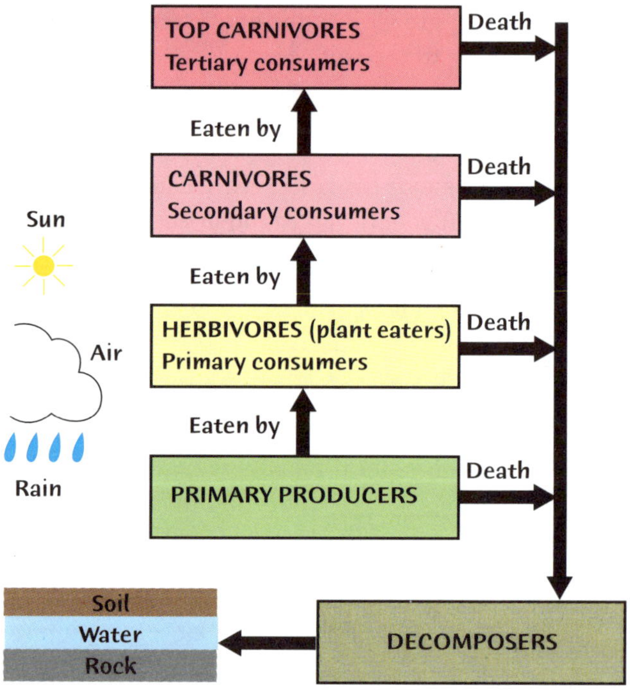

Nutrient recycling

- Nutrients move through ecosystems and provide plants and animals with food.
- Rocks weather down to form soil.
- Nutrients (and water) in the soil enter living plants via the root system.
- Decomposers return nutrients to the ecosystem.

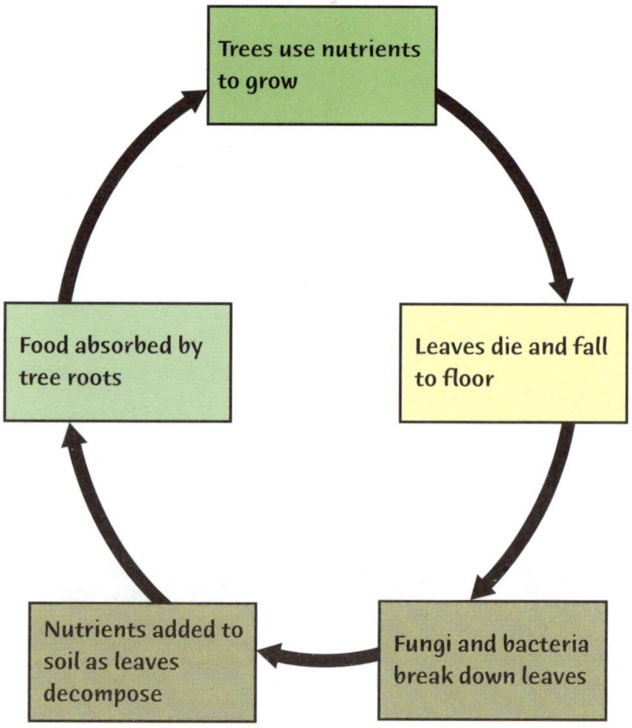

THE CRASH COURSE

The world's major ecosystems

The world can be divided into a number of major ecosystems called **biomes**. Each biome has distinctive characteristics, influenced mainly by the climate.

Six selected world biomes

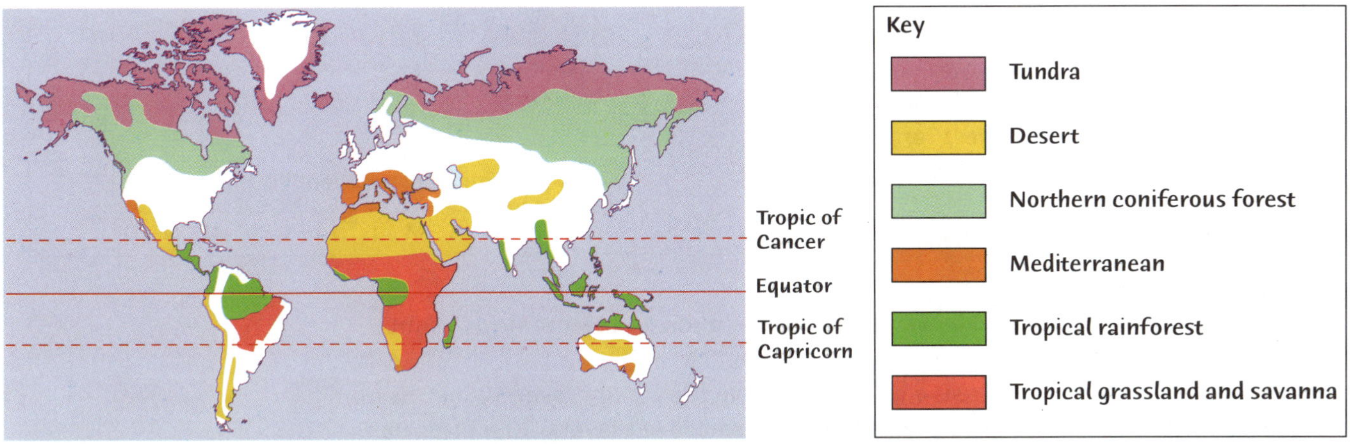

EXAMPLES

Example 1: The savanna ecosystem

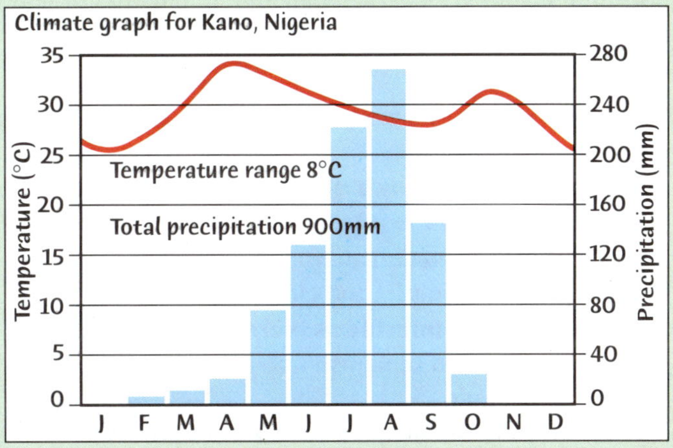

Location (see map above):
A zone from 5° to 15° north and south of the Equator, between the world's rainforests and hot deserts

Climate:
- high temperatures, over 25°C all year
- low temperature range, less than 10°C
- dry season, 7–10 months; rainfall 0–20mm per month
- hot, wet season, over 30°C and above 200mm rainfall per month
- thunderstorms common in rainy season

Vegetation:
- a mixture of scattered trees, bushes and grasses – mainly acacia and baobab trees and thorn bushes
- vegetation has to be able to survive high temperatures and a long dry season
- adaptations include small waxy leaves, long roots, thorns and the ability to store water
- trees and shrubs are deciduous, losing their leaves to conserve water
- grasses grow tall quickly when there is rain, dying back to the roots in the dry season
- at the drier edges of savanna, which borders desert, there are fewer trees and bushes, and shorter grass
- towards the wetter borders, closer to the rainforests, there is much more vegetation

Land use:
Despite water shortages, savanna areas are important farming areas, although over-grazing and unreliable rainfall are major problems in many areas. Trees are cut down for fuel, removing shade and also leading to soil erosion.

Typical savanna landscape – acacia trees and short grasses

The Crash Course

Examples

Example 2: The rainforest ecosystem

Location (see map on page 41):
5° either side of the Equator

Climate:
- hot, over 25°C, and humid all year
- very low temperature range: 1° or 2°C
- wet all year with rainfall totals often over 2000mm per year
- heavy (convectional) rainfall most afternoons

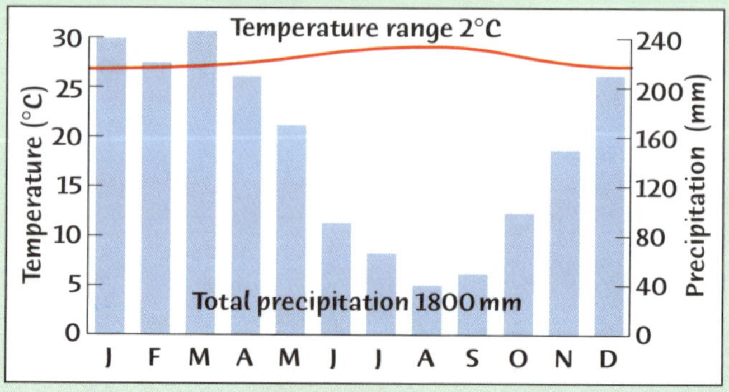

Vegetation:
- the richest and most varied biome in the world – animals, insects, birds, plants
- distinctive layers with characteristic vegetation
- emergents – isolated, tall trees e.g. kapoks, above the canopy; over 50 metres up
- canopy – a sea of leaves, blocking out the sun from lower layers; 30 metres up
- under canopy – mainly bare tree trunks and lianas (vines); dark; 20 metres up
- shrub layer – shrubs, saplings, up to 10 metres above ground level
- ground level – very dark, damp and humid, lots of leaf litter, dense undergrowth and buttress roots of large trees

Deforestation:
The world's rainforests are being cut down at an alarming rate, for all sorts of reasons. This results in a huge range of problems:
- soil, once protected by a thick canopy of vegetation and held together by tree roots, is easily eroded
- with no vegetation to intercept the heavy rainfall, surface run-off and flooding increase
- on a global scale widescale forest destruction increases CO_2 levels … and global warming
- the soil becomes infertile as there is no leaf litter to decompose and replenish nutrients, making farming difficult
- valuable, slow-growing hardwood trees are destroyed
- mining for valuable minerals can cause pollution
- wildlife habitats are destroyed and many animals move away or are killed
- large areas may be flooded by damming rivers to build hydro-electric power stations and reservoirs
- valuable medicines may be lost – over half these today come from rainforests
- forest Indians lose their traditional lands or are forced to move elsewhere

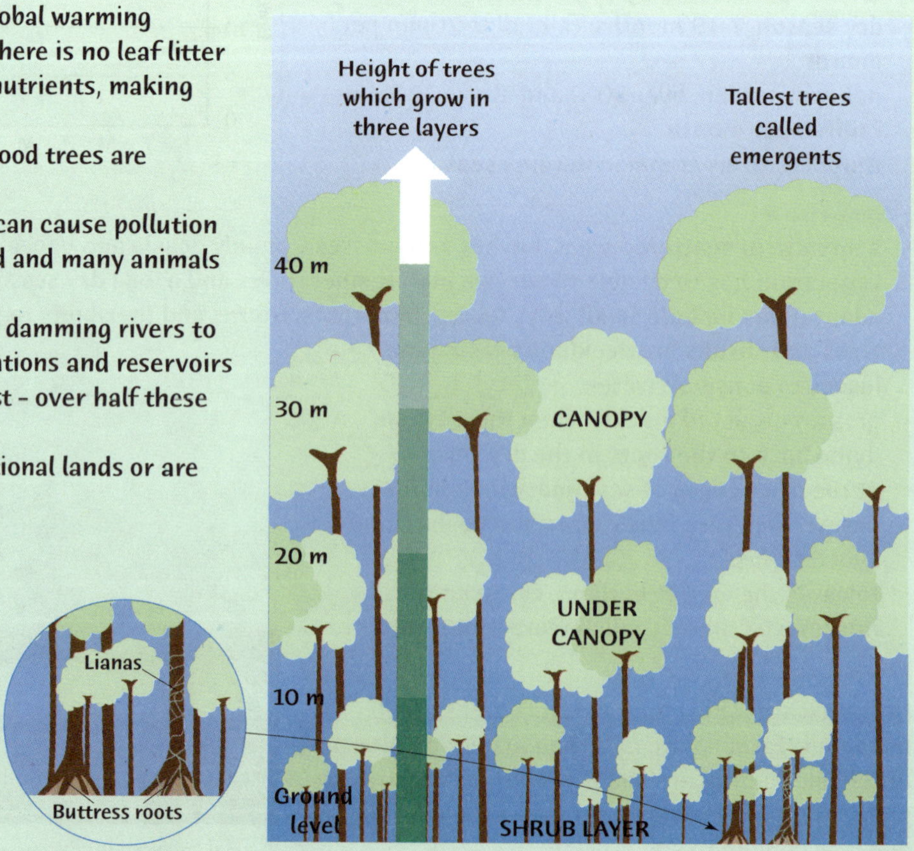

CASE STUDIES

EXAMPLES

Commonly used case studies:

Rainforests:	Amazon (Brazil); Sarawak (Borneo); Indonesia; Malaysia; Cameroons
Coniferous forest:	Caledonian Forest (Scotland); Finland; Canada; Western Siberia
Deciduous forest:	UK local study; New Forest (Hampshire)
Savanna:	Kenya; Nigeria; North Australia; Tanzania; South Africa
Mediterranean:	Mediterranean Europe e.g. southern Italy / Spain; California (USA)
Hot desert:	Mojave (USA); Namib and Kalahari (Namibia);
Wetlands:	UK local study; Florida (USA)

Remember that ecosystems vary in size – you may have studied small local ecosystems and/or large biomes.

Sample case study: Amazon rainforest

Location:

North-west Brazil, stretching along the Equator from the Andes to the Atlantic Ocean

Description:

- largest rainforest in the world
- ten times the size of the UK
- drained by the second longest river in the world, the Amazon
- hot (over 27°C) and humid all year
- in places, over 2000mm of rain per year
- contains about 80% of all known species of plants, animals and insects

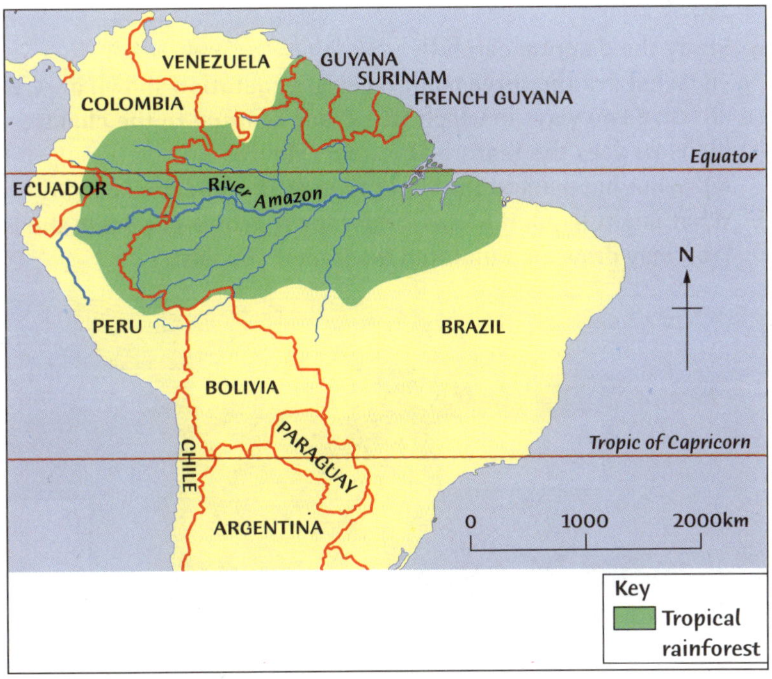

Issues:

Almost half of the original area of the Amazon rainforest has been cut down as the Brazilians – and others – have tried to exploit the land and its resources for:

- cattle farming – large areas have been cleared and burnt for ranching, usually to produce meat for export
- subsistence farming – native Indian tribes and people from other areas (encouraged to move from urban areas by the government) clear small areas to produce crops
- plantation agriculture – natural forest has been cleared to grow plantation crops e.g. rubber, cacao and coffee for export
- logging – valuable hardwoods e.g. mahogany have been cut down, fetching high prices in MEDCs but leading to widespread forest destruction
- mining – the forest contains a great range of resources e.g. bauxite, gold, iron ore and mercury. Areas like Trombetas and Carajas have been cleared for mining minerals, mainly for export
- hydro-electric power – power stations have been built e.g. Tucurai, flooding large areas of the forest
- transport – the construction of the Trans-Amazonian Highway (part of the Pan American Highway) has opened up the forest, making remote areas accessible

Many people think that Brazil should stop destroying the Amazon rainforest, but Brazil is an LEDC trying to pay off large debts and develop its own industries by developing its own resources.

EXAM PRACTICE

Question A

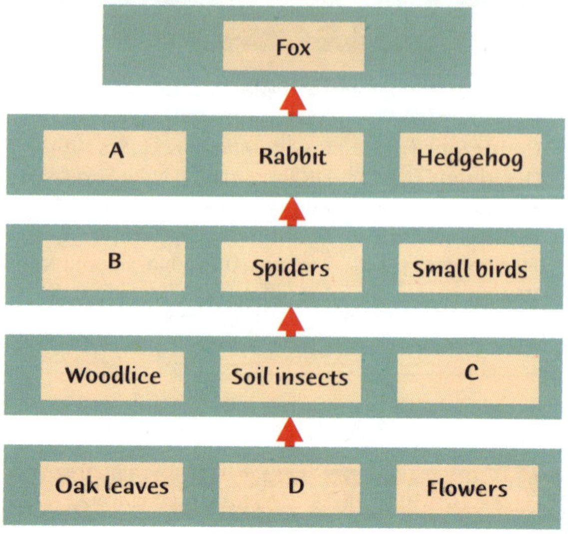

a Study the food chain diagram (part of a deciduous woodland ecosystem). Complete spaces A, B, C and D with the correct words from the following:

owl caterpillar
mice grasses

b Name: i) a primary producer ii) a secondary consumer.

c Explain in detail what might happen to the food chain if all the caterpillars died.

d i) What is biomass?
ii) Which part of the food chain has the highest amount of biomass?

e Decomposers are vital parts of an ecosystem / food chain. Explain why.

Question B

a Study the diagram carefully.
i) What are the three main types of vegetation shown at A, B and C?
ii) Give two ways in which tree B has adapted to the climate.
iii) How does the feature at A help wildlife?
iv) What happens to the vegetation at D in the dry season?

b What might happen to this landscape if people cut down the trees and started to rear animals? (You may draw an annotated diagram if you wish.)

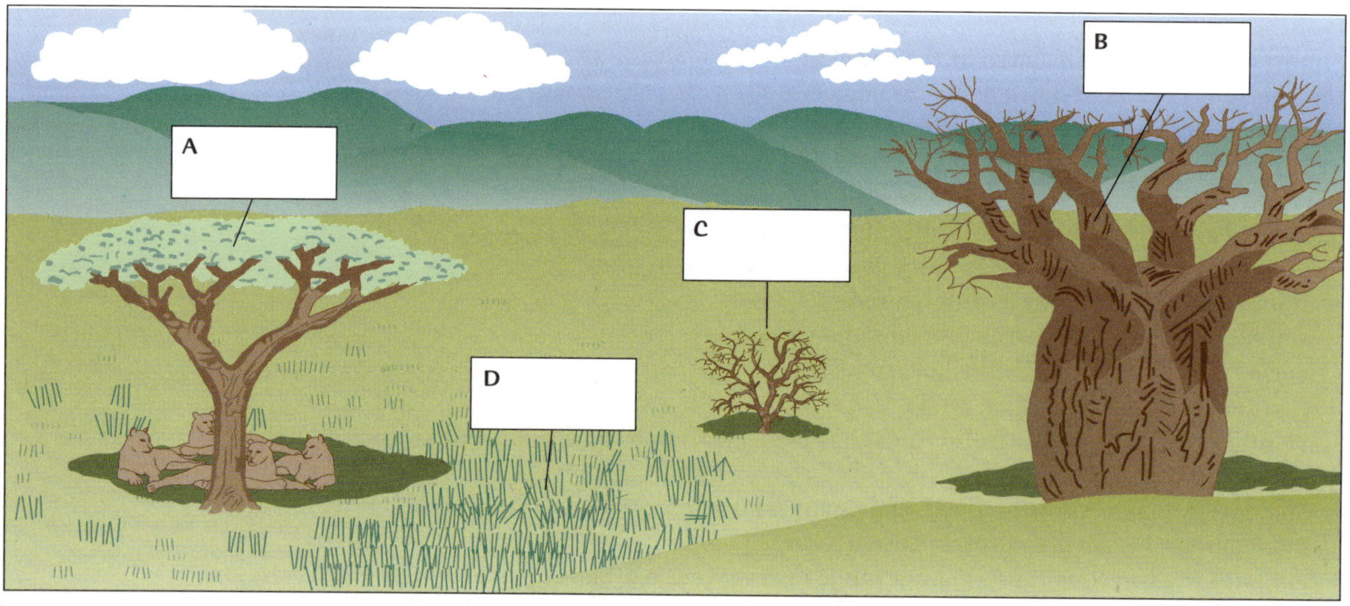

Question C: case study question

a For a named ecosystem you have studied, and which people have changed, name and locate the ecosystem.

b Describe how the ecosystem has been changed.

c Explain the good and bad effects of the changes.

Revision Unit 1

Population

The Check up

1 The world's population is not __________ distributed – some areas are empty and others are crowded.

2 The most crowded areas of the world are called __________ populated.

3 Population density is the __________ of people in a given __________ e.g. per km^2.

4 World population is expected to __________ in the next 50 years, from 6 billion to __________ billion.

5 The model used to show world population changes is called the __________ Transition Model. It shows the __________ stages of population growth.

6 __________ rate is the number of __________ born each year per __________ of the total population.

7 Look at diagram (a). Use it to help you complete the following statement:

Growth rate or natural increase is __________ rate minus __________ rate.

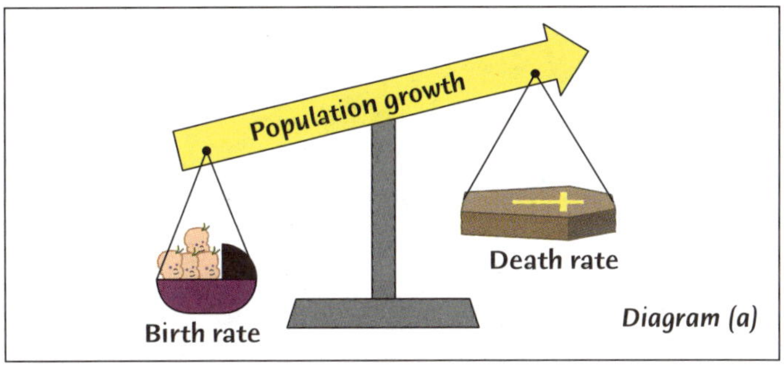

Diagram (a)

8 Growth rate is generally higher in __________ countries compared to __________ countries.

9 Good __________ control provision is essential to help reduce rapid population growth.

10 Better __________ care means that many people are now __________ longer.

11 Look at diagram (b), a population pyramid. Complete the labels A, B and C.

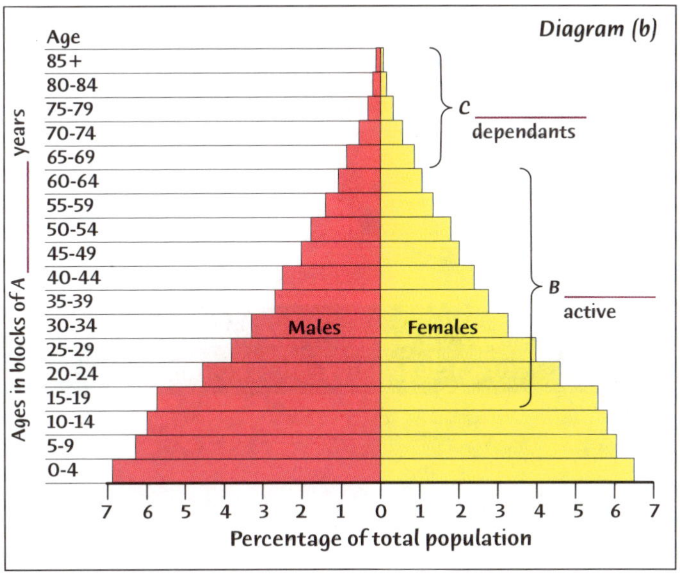

Diagram (b)

12 The population pyramid in diagram (b) is typical of a __________ country because it is __________ at the base and __________ at the top.

13 __________ countries are more likely to have __________ shaped graphs, with fairly even numbers in most age groups.

14 Studying population structure can help us plan for the future e.g. building enough schools if there are many __________ dependants, providing adequate health provision if there are many __________ dependants.

15 The __________ of people from place to place is called migration.

16 When people move for a short period of time it is called __________ or __________ migration.

17 __________ to __________ migration is when people move from the countryside to live in towns or cities.

18 Many people in MEDCs are starting to move out of cities. This process is called __________ urbanization.

19 People may move away from an area because of negative reasons – these are called __________ factors.

20 If people are attracted to a new area because of, for example, better facilities, these are called __________ factors.

Mark yourself *(see answers on page 99)*

- More than 12? Do the Exam Practice – or go on to another unit.
- Less than 12? Do the Crash Course on the next page.

____ / 20

The Crash Course

Population distribution - where people live

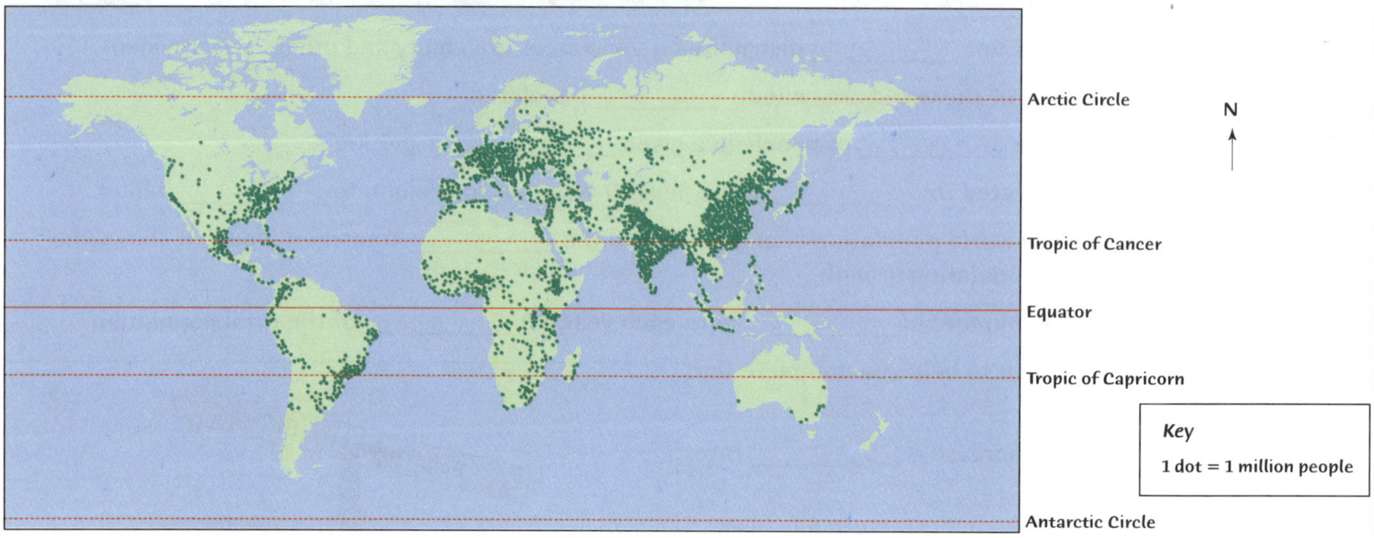

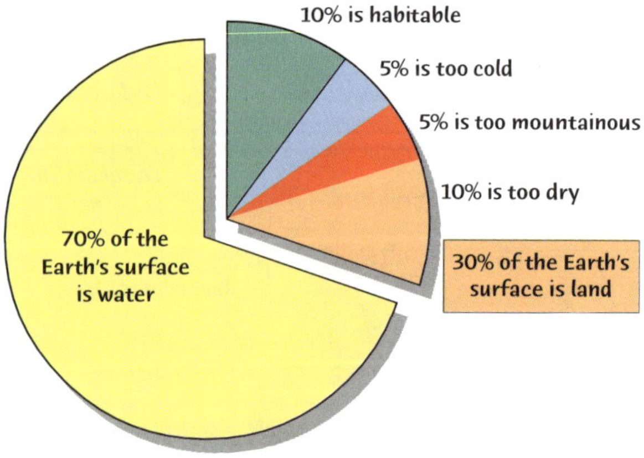

- The world's population of 6 billion people is not evenly spread.
- Most people live in just one third of the land area.
- Population density is the number of people living in a given area - this may be given as a number per km² or as a national total (total population divided by the total area).
- Some areas are densely populated = a high number of people per km².
- Some areas are sparsely populated = few people per km².
- There are usually good geographical reasons for this pattern of distribution (see pie chart).

Tip: Maps on a world scale hide regional and local differences - the figures shown are averages for each country. Most countries will have both sparsely and densely populated areas.

Changing world population - a model for population growth

- The model here is called the **Demographic Transition Model**.
- Geographical models try to simplify complex world situations.
- It shows four stages of population change (see page 47).
- The total world population is increasing rapidly.
- There are six times as many people in the world today compared with 1800.
- The total is expected to double in the next 50 years.
- More people are being born each year than the number of people dying.
- People are living longer e.g. because of better medical facilities and better diets.
- Like population distribution, population growth is uneven.

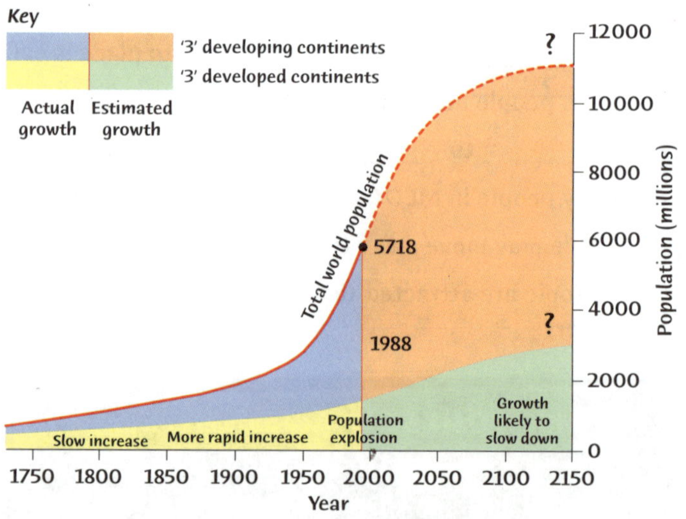

The Crash Course

Population growth

- **Birth rate** is the number of babies born each year per 1000 of the total population.
- **Death rate** is the number of deaths each year per 1000 of the total population.
- **Natural increase / growth rate** is the birth rate minus the death rate (given as a percentage).
- The natural increase or growth rate across the world varies considerably.
- In most MEDCs, growth is either low or static – in some it is even expected to fall in the next 20 years.
- Birth control / contraception in MEDCs is widely available – many people choose to have fewer children.
- In LEDCs the natural increase is generally much higher.
- Families in LEDCs are generally larger, to earn money and help work the land.
- Birth control in LEDCs can be expensive or against religious beliefs.
- Infant mortality is much higher in LEDCs which gives rise to higher birth rates because many children do not survive.

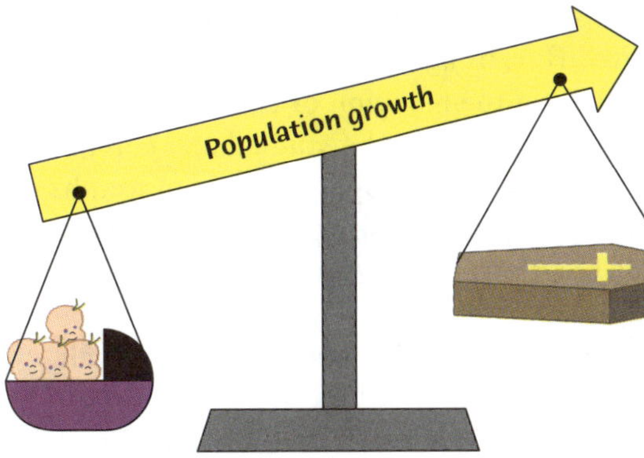

Birth rate more than death rate

Stages of population growth

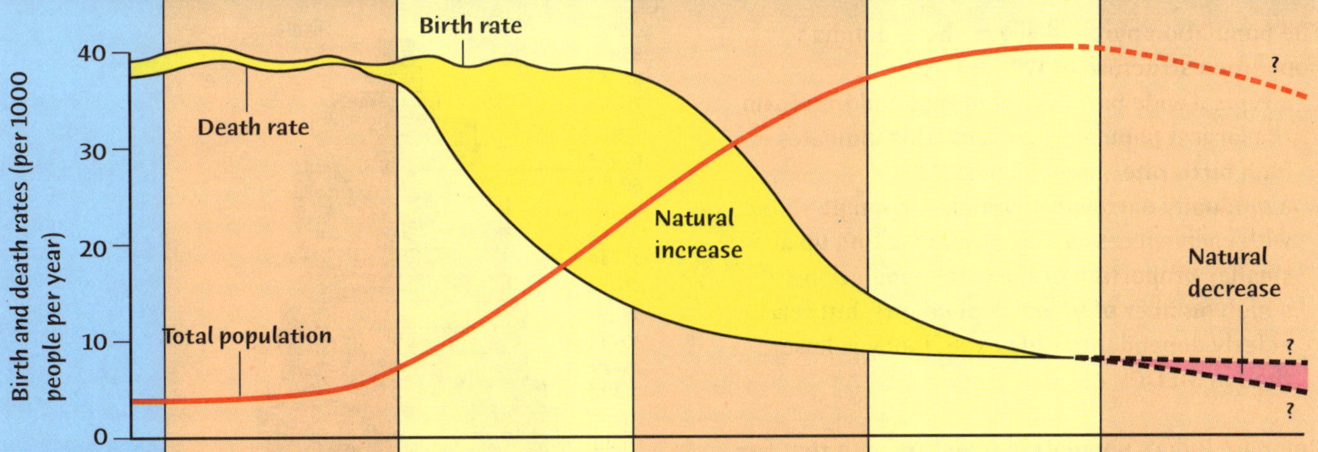

Stage	1 High stationary	2 Early expanding	3 Late expanding	4 Low stationary	5? Declining?
Examples	A few remote groups	Egypt, Kenya, India	Brazil	USA, Japan, France, UK	Germany
Birth rate	High	High	Falling	Low	Very low
Death rate	High	Falls rapidly	Falls more slowly	Low	Low
Natural	Stable or increase	Rapid increase	Increase slows down	Stable or slow	Slow decrease

The model on page 46 shows overall world population growth. Population growth rates of individual countries vary greatly. The model above shows that there is a distinct pattern to growth. Countries develop through four, perhaps five, main stages as the population grows and the economy develops.

The Crash Course

Population structure

Population structure is the numbers of males and females in different age groups in a given population e.g. a town, region or country. So that population structures can be studied and compared, they are usually drawn as special types of graphs called **population** or **age-sex pyramids**. These are really two bar charts drawn back to back. They show:

- the percentage of **males** and **females** in each age group
- the percentage of males on the left and females on the right
- the percentage of people in each **five year age group**
- the percentage of economically-active and economically-dependent people.

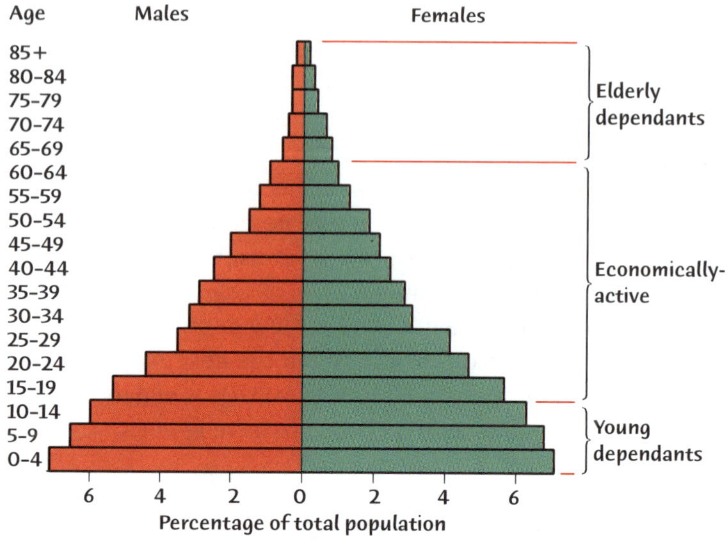

India, 1991

Population pyramids can be used to predict population changes, and to plan for the future. For example, it can be seen how many schools may be needed; if there is an imbalance between males and females; and the proportion of elderly people in the population who may need additional health service provision. When compared over time or between different countries they can be an indicator of development.

Comparing development - MEDCs and LEDCs

The population pyramid above showed India's population structure in 1991. It has:

- a typical wide base, showing that children form the largest population groups. This indicates a high birth rate, common in LEDCs;
- a gradually narrowing pyramid / triangle shape with each successive age group making up a smaller proportion of the total population;
- a high number of young dependants, but few elderly dependants - life expectancy is lower than in MEDCs.

Compare India's population structure with that for the United Kingdom in 1991. Note in the UK's pyramid:

- there is no real pyramid shape - the percentage for each age group only varies slightly;
- the 'rectangular' shape of a population reflects a low death rate / low infant mortality / high life expectancy;
- the 'bulge' or widest part of the graph occurs between ages 15 and 24;
- there are far more elderly females than males;
- unlike India, the two dependant groups - the young and the elderly - are similar in number.

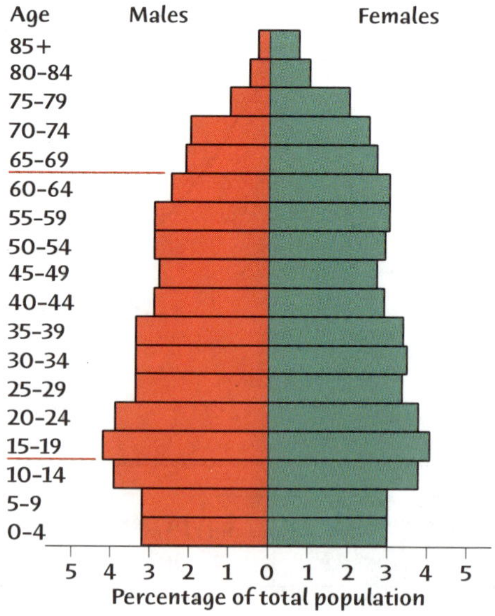

UK, 1991

Both graphs give useful information about each country, especially the proportion of the population who are economically active, i.e. working, earning money and paying taxes. The economically-active population supports the proportion of the population who are classed as dependant.

The Crash Course

Migration

People seldom live in the same place for their whole lives. Moving from one place to another is called migration. There are many different types of migration.

Key facts to learn: types of migration

- **immigration**: moving into another country
- **emigration**: moving out of a country
- **international**: moving from one country to another
- **voluntary**: moving by choice
- **forced**: having to move because of war, political asylum, natural disaster. These types of migrants are called **refugees**
- **temporary/seasonal**: moving for a short period of time e.g. to find work (harvesting crops)
- **rural to urban**: people moving from the countryside to live in towns and cities; an increasing problem in LEDCs. This leads to increasing **urbanization** i.e. the growth and spread of urban areas. In many large cities in LEDCs it has led to the growth of shanty towns or squatter settlements on the edges of the city (see p 57). Problems can also be created for the rural area left behind by the migrants – less workers, services decline.
- **urban to rural**: in many MEDCs people are moving out of crowded urban areas into rural areas. This is called counter-urbanization. This can cause problems as new housing may have to be built in greenfield locations and this may create conflict between new and existing residents. Urban areas may also start to decline as people move away – shops close, buildings become derelict.

Why people migrate

People may migrate because of poor conditions where they live – these negative reasons are called **push factors**. Some examples of push factors are:

- lack of jobs or job opportunities
- poor pay
- poor services (health, education)
- isolated / few facilities (leisure, transport)
- climatic hazard e.g. drought leading to crop failure
- other hazards e.g. civil war
- poor quality of life.

Others move because they are attracted by conditions elsewhere – these are positive reasons called **pull factors**. Some examples of pull factors are:

- job opportunities
- chance to earn more money
- range of services (hospitals, schools, cinemas)
- more to do e.g. entertainment facilities
- less likelihood of natural hazards
- better quality of life.

Sometimes people migrate because of a combination of push-pull factors; sometimes they migrate for a very specific reason.

CASE STUDIES

EXAMPLES

Commonly used case studies:
Population studies:
MEDCs: UK (various); France; Italy; Japan; USA
LEDCs: India; Bangladesh; Kenya; South Africa; Brazil

Migration: Ugandan Asians to the UK; Turks to Germany; refugees from former Yugoslavia to other EU countries; Mexicans to USA; Rwandans to Tanzania

Sample case study: Mexico to USA migration

Location:

Along the 2000km long border between southern USA (California, Arizona, New Mexico and Texas) and Mexico

	USA	Mexico
GNP per head	$26 980	$3 320
Life expectancy	73 (m) 80 (f)	67 (m) 74 (f)
Unemployment rate	5.6%	28%
Infant mortality rate (per 1000)	7	24
Birth rate (per 1000)	15	26
Adult literacy	99%	87%

Datafile on the USA and Mexico

The Mexican/US border

Description:

- there are huge differences in wealth and quality of life between the USA and Mexico (see datafile)
- approximately twelve million Mexicans have legally emigrated to the USA
- there are probably four to five million Mexicans living illegally in the USA
- early migration was encouraged – workers were needed for a range of (usually low-paid) jobs in the southern states
- some migrants are seasonal or temporary e.g. harvesting summer crops, others try to get jobs in cities like Los Angeles
- most migrants are looking for paid work and a better quality of life than in Mexico
- in the 1970s and 1980s the economic recession made jobs harder to find and resentment grew against Mexican immigrants who had imported their own culture via shops and restaurants
- an estimated million or more Mexicans try to enter the USA illegally every year – in 1995, 800 000 were caught and sent back
- the majority of those trying to enter the USA are young men – this causes problems in Mexico where children and old men and women make up the majority of the population

Migration from Mexico to the USA has been happening for over a hundred years. Migrants were needed for industry to develop in states like California. However, in times of recession in the USA, migrants are seen as a threat to jobs and a drain on resources. The border between the two countries is very long and almost impossible to fully police, so illegal migration is not difficult. Mexicans form the majority of the Hispanics (Spanish speakers) who account for almost 10% of the USA's total population today.

Exam Practice

Question A

a What is meant by: i) densely populated ii) sparsely populated ?
b Study the map. Name: i) three densely populated areas ii) three sparsely populated areas.
c The UK has a population density of 234 per km^2. Why is this figure misleading?
d Write a paragraph describing population distribution in the UK.
e Explain why some areas of the UK are densely populated whilst others are sparsely populated.

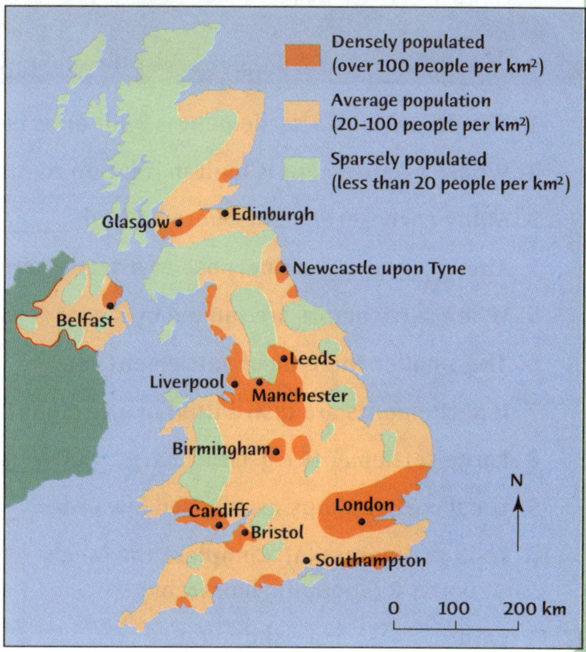

Question B

Look at the two population pyramids, A and B.
a i) What percentage of the population in A is under 15?
ii) What percentage is over 60?
b Between which ages are people classed as economically active?
c In B, which two ages groups have most people?
d Describe the shapes of A and B.
e i) Which pyramid shows an LEDC and which shows an MEDC?
ii) Gives reasons for your answers to (i).
f How and why might an event like war 20 years ago change the shape of pyramid B? (You may choose to draw an annotated diagram for your answer.)

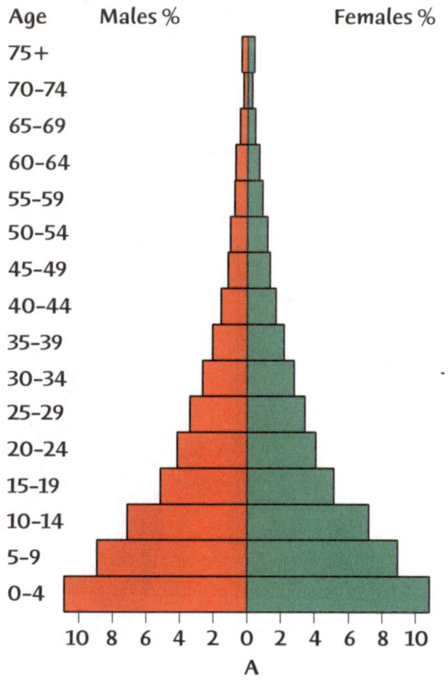

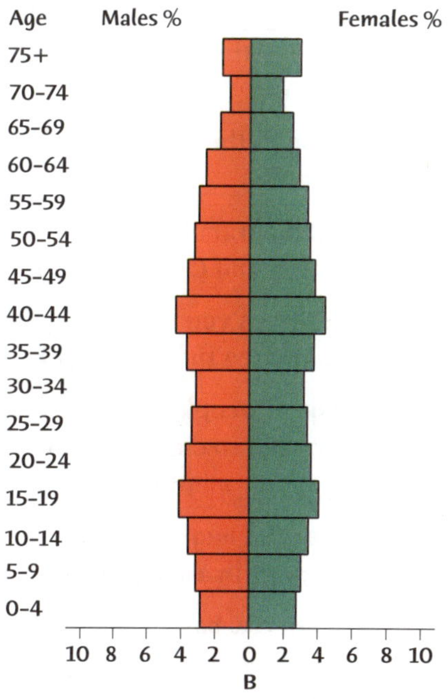

Question C: case study question

For a named example of migration you have studied:
a Name the area from which and to which migration took place.
b Under the headings **push factors** and **pull factors** list the reasons for the migration.
c What disadvantages did the migrants experience?
d How were the areas the migrants left affected?

Revision Unit 8

Settlement

The Check up

1 The actual land a settlement is built on is called the __________.

2 Where a settlement is built in relation to other places is called its __________.

3 Many early settlements were built on __________ tops so that they could be easily defended.

4 The __________ or purpose of a settlement is what it is used or built for.

5 When settlements are ranked by size it is called a __________.

6 The smaller the type of settlement, the __________ of them there are.

7 If settlements are distributed in a straight line the pattern is known as __________.

8 Large settlements e.g. cities have __________ services than villages.

9 Local corner shops usually sell low order __________ goods.

10 Look at diagram (a). Complete the labels A, B and C (shopping hierarchy).

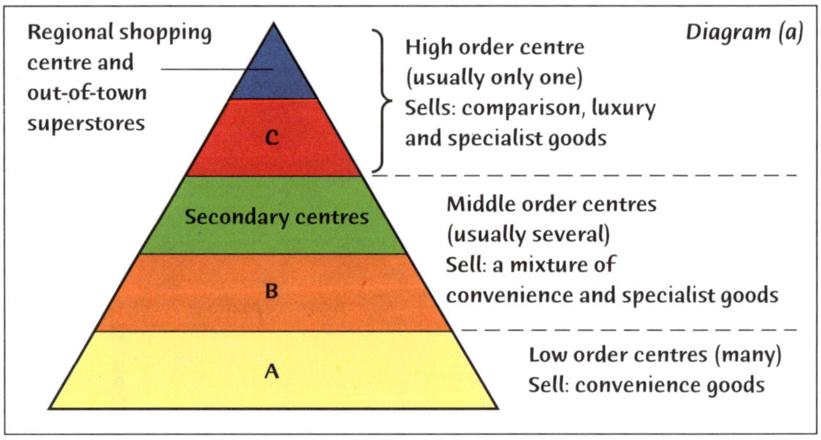

11 Large __________ shopping developments are found on the edges of urban areas.

12 In the UK __________ out of ten people live in urban areas.

13 Urban __________ try to show in a simple way the way towns and cities have grown.

14 The zone in the middle of an urban area is called the __________.

15 Look at diagram (b). Name the land use found in zones A, B and C (Hoyt model).

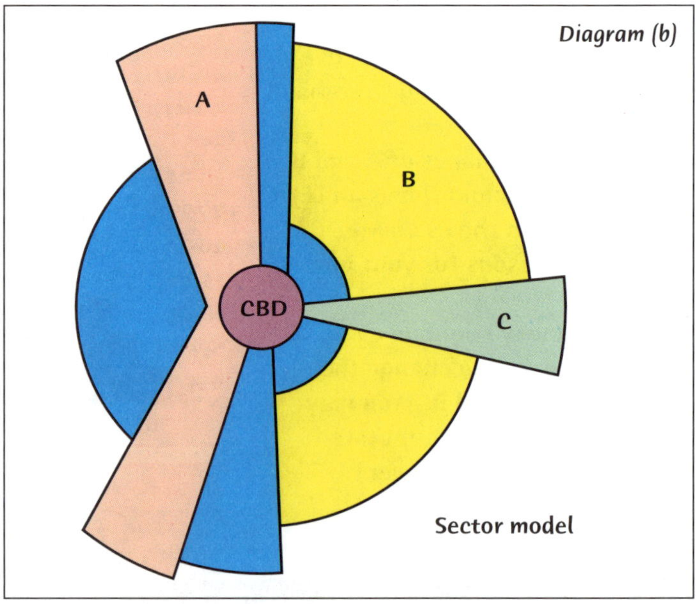

16 In the UK, as you move away from the middle of an urban area the buildings become __________.

17 Millions of people have moved from the countryside to cities in LEDCs – this is called __________ migration.

18 Many of the migrants to LEDCs settle on the edge of the cities in areas called __________.

19 There are few basic amenities e.g. __________, sewage disposal and __________ in this outer zone in LEDCs.

20 In LEDCs these areas are often improved by the residents themselves by __________ schemes.

Mark yourself *(see answers on page 99)*

- More than 12? Do the Exam Practice – or go on to another unit.
- Less than 12? Do the Crash Course on the next page.

/ 20

THE CRASH COURSE

Settlements

Settlements are places where people live. They range in size from isolated farms to large cities or conurbations. The actual place where a settlement is built is called its **site**. Where it is in relation to other settlements and nearby physical features and resources is called its **situation**.

Choosing a site

Most settlements in the UK are hundreds of years old, so the reasons for choosing the original site were usually based on the needs of the time, rather than those of today. These needs included:

- a good **water supply** e.g. spring, stream or river
- **dry land** to build on – not on marshy land or too close to a river e.g. above flood level
- somewhere easy to **defend** from attack e.g. on top of a hill
- availability of **building materials** e.g. wood, stone / rock
- **food supply** – suitable land to farm or natural resources to exploit
- **shelter** from extreme weather e.g. lower slopes of a hill or valley
- **fuel supply** e.g. wood to burn for cooking and heating
- easy **access** to other places e.g. along rivers or valley floors.

The middle and lower courses of river valleys were often chosen since they could provide most of the basic needs listed above, especially at points where the river could be easily crossed and a bridge built. Settlements were often built on the lower valley slopes.

Sketch maps showing site and situation

- Map (a) is a sketch map showing the site of Carlisle.
- Carlisle has developed around a bridging point on the River Eden.
- This is a good example of an annotated sketch map.
- Annotations are detailed labels which help explain what is there – and why.
- A map showing the situation of Carlisle would cover a larger area, at a smaller scale. It would show links to other settlements and other features.

The ability to draw sketch maps showing site, situation or other specific information is a very important geographical skill. Adding labels or annotations often saves time in writing much longer answers.

Map (a)

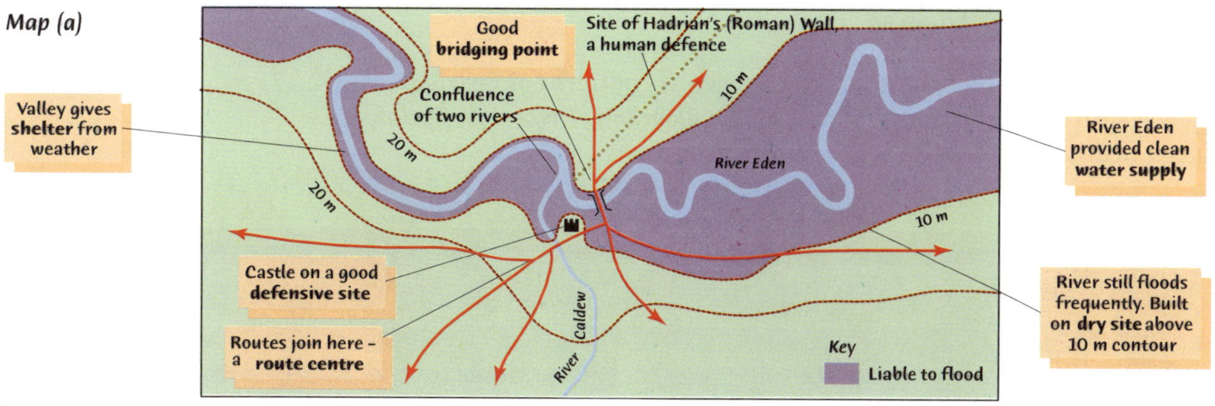

The Crash Course

Functions of settlements

The function of a settlement is what it is used for – its purpose or why it was built. A settlement may have more than one function, and may actually change its function e.g. a mining town where the mines have closed down. Some of the main functions are:

- port: trading / imports and exports via ships and boats e.g. on a river estuary or coastal location
- market: acts as a buying and selling point for local goods and produce e.g. centre of a farming area
- industrial: manufacturing goods, factories and raw materials e.g. coal mining, textiles, food processing, car assembly
- holiday resort: attracts visitors / tourists e.g. seaside town, area of outstanding natural beauty
- educational: concentration of universities and colleges
- commuter: most inhabitants travel out every day to and from work
- administrative: e.g. county or regional town, capital city.

Blackpool – a seaside town

Settlement types and hierarchies

- Settlements can be classified according to size.
- They are ranked in order, smallest first, giving the distinctive settlement hierarchy shown in the diagram.
- As you move up the hierarchy, the size of the settlement increases, as does the number and range of buildings, facilities, services and people in them.
- The total number of each type of settlement decreases as you move up the hierarchy.
- The range of services found in settlements e.g. shops, education, leisure and health services, can also be placed in hierarchies.

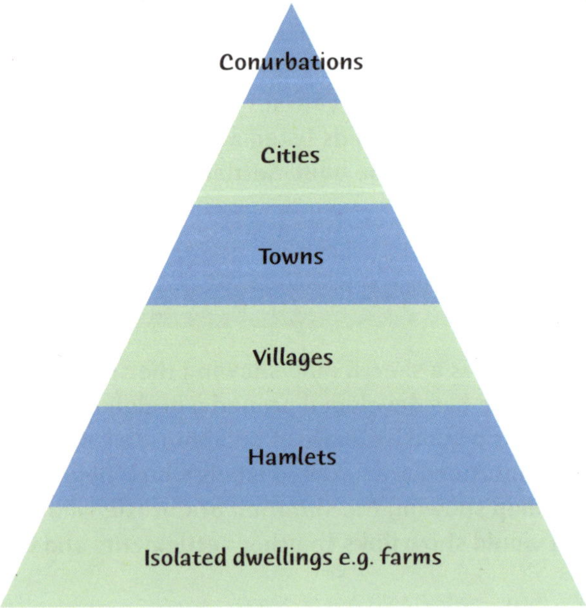

Distribution of settlements

If you look at settlements over a large area you can identify patterns of distribution. These fall broadly into the four patterns shown by the diagrams. These descriptions can also be used to describe the pattern of services in an area.

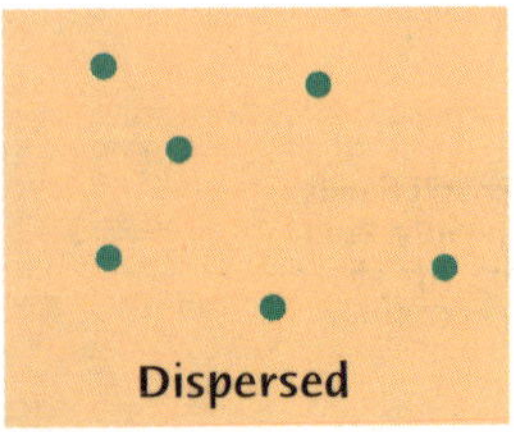

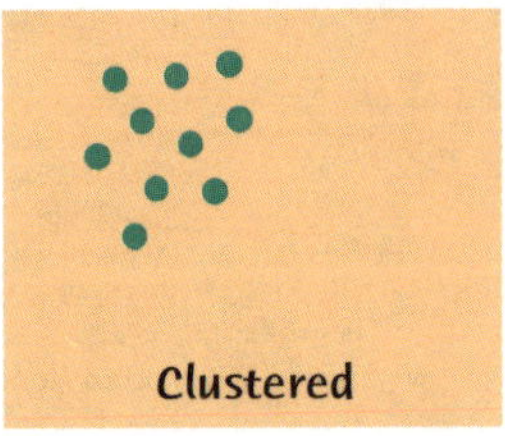

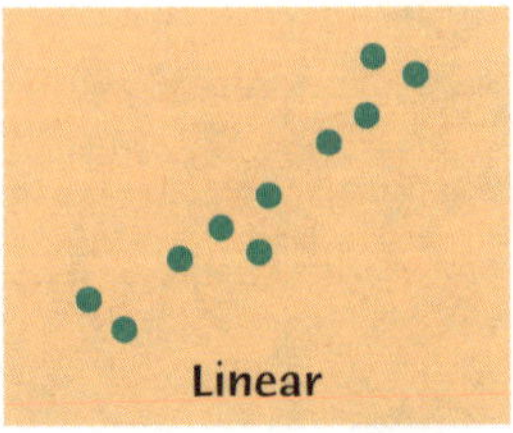

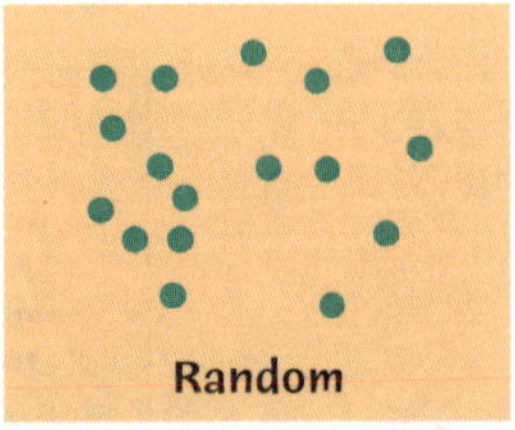

The Crash Course

Settlement: provision of services

Large settlements also act as service centres – places where people will travel to shop and use the range of services on offer e.g. health centres, libraries, financial services. Larger settlements will provide a much greater range of services than smaller ones. This is particularly true of shopping services. Different types of shops / shopping centres can be arranged in a shopping hierarchy like the one on the right.

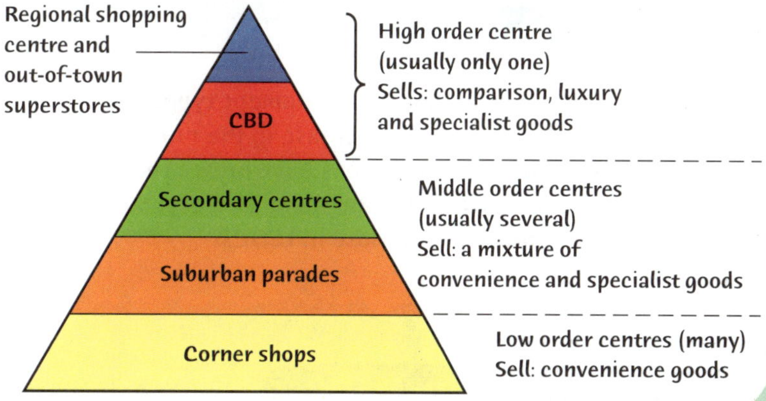

Characteristics of different shopping centres

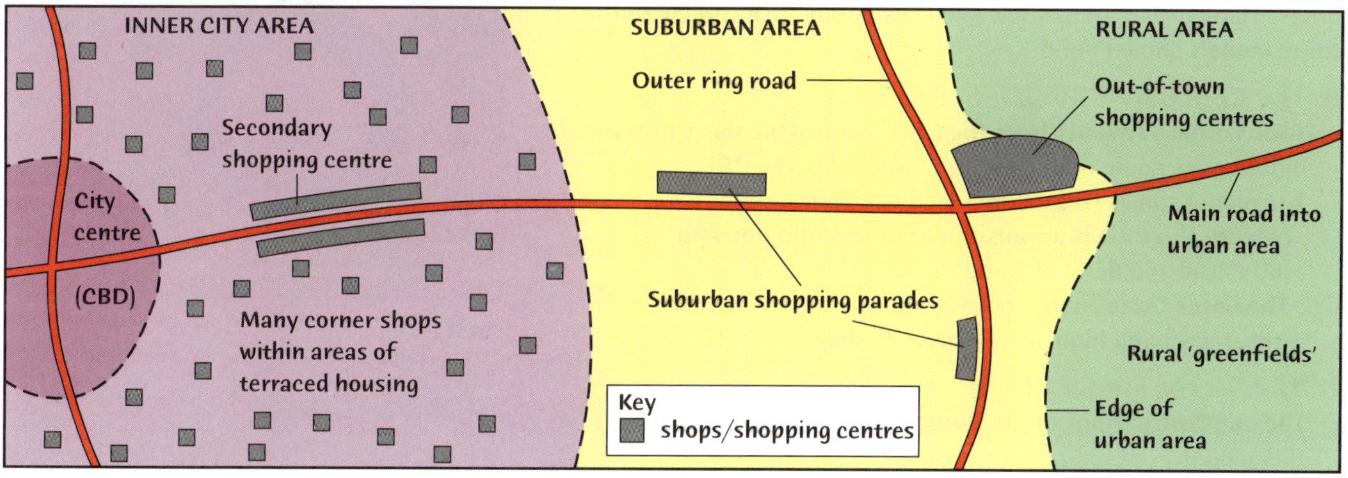

Corner shops:

- sell a range of convenience goods e.g. milk, bread, stamps, newspapers and general groceries
- serve local residential areas – in walking distance
- usually open long hours
- often on street corners (hence the name) converted from houses.

Suburban shopping parades:

- usually eight to ten shops grouped together
- a mixture of general and specialist shops e.g. bakers, butchers, small chain store, fast food outlets
- easy access for local people, some car parking space
- purpose-built to serve what were then new residential areas / estates.

Secondary shopping centres:

- along main roads leading to CBD – number of shops vary from around 20 to over 100
- general and specialist stores, banks, restaurants, petrol stations, small supermarkets
- originally accessible and easy to park, but more difficult today with urban parking restrictions
- used by locals and people passing through / not wanting to go into CBD.

CBD (city centre):

- great range of shopping goods and services
- town or city centre – main routes meet here, although accessibility can be difficult (many CBDs are now pedestrianized)
- large chain and department stores, jewellers, restaurants and comparison goods
- people may travel long distances to shop for specialist / expensive goods, and to have a wider choice of goods at more competitive prices.

Out-of-town developments:

- edge of urban areas, on or near major roads or motorway junctions for accessibility (for customers and for goods coming in)
- large areas of cheap land – big stores with plenty of parking space
- supermarkets and big chain stores – great range, cheap / competitive prices
- modern, under cover and usually provide leisure facilities, child-care and a great range of food and drink outlets.

THE CRASH COURSE

Urban growth

Although there is a wide range of settlement types in the UK, nine out of every ten people live in urban areas. The increase in the numbers living in towns and cities is called **urbanization**. As urban areas have grown they have developed common characteristics in terms of structure and land use. Such patterns vary greatly between cities in MEDCs and those in LEDCs.

Land use in cities in MEDCs

The structure of urban areas in MEDCs tends to follow similar shapes and patterns. These are usually shown as land use models – simplified maps showing different land use zones within urban areas. Many urban areas show patterns of land use similar to the three models shown here.

A The Concentric Model (Burgess)

- Towns grew outwards from the CBD (central business district).
- Buildings become newer moving away from the CBD.
- The model shows five distinct land use zones:
  Next to the CBD is a transitional zone of old housing, industrial and derelict land.
  The outer three zones are residential – low (old), medium (inter-war) and high-class (new) housing.

B The Sector Model (Hoyt)

- The concentric rings of the Burgess Model form the basis for this model.
- As transport developed in the twentieth century, growth was outwards in sectors or wedges following canal, rail and road links.
- Other sectors included industry with low-cost housing for workers.
- Middle-class and high-class housing also developed along transport routes in other parts of the city.

C The Multiple Nuclei Model (Harris and Ullman)

- Both the first two models are fairly simple – in reality the land use pattern in urban areas today is more complex.
- Urban growth is rapid, taking in outlying small towns and villages.
- Apart from the main CBD, there are a number of other, smaller centres.
- Land use is still zoned but there is more variety.

A

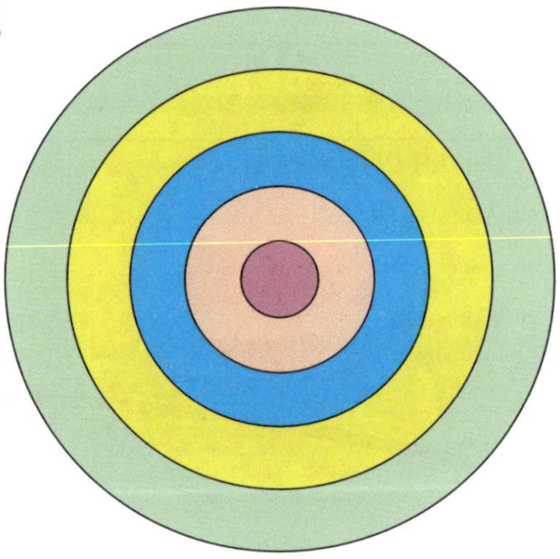

B

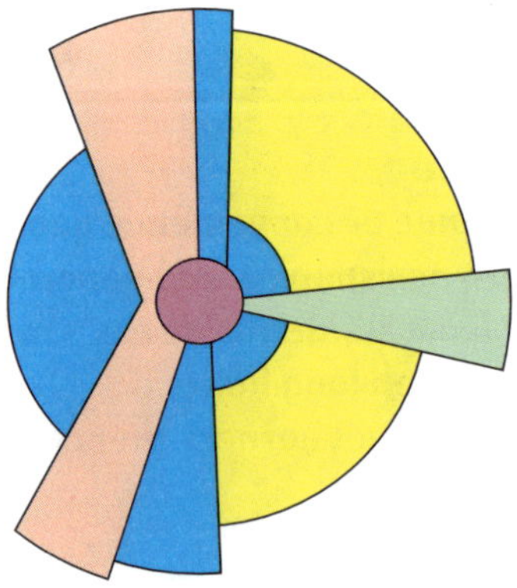

C

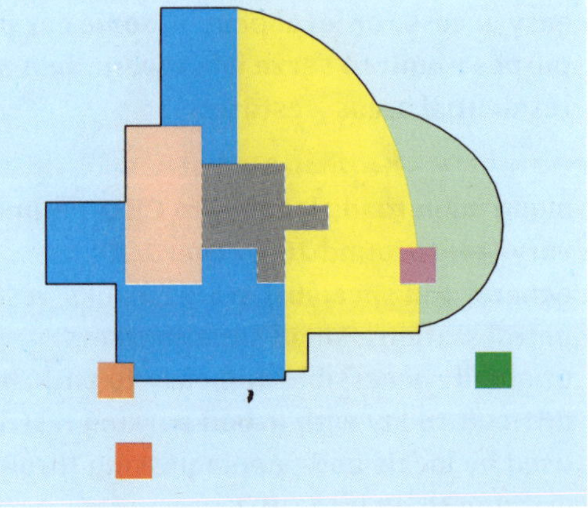

- Central Business District (CBD)
- Wholesale light manufacturing (transitional)
- Low-class residential (old inner city areas)
- Medium-class residential (inter-war areas)
- High-class residential (modern suburbs)
- Heavy manufacturing
- Outlying business district
- Residential suburb
- Industrial suburb
- Commuter zone (Building age decreases outwards)

The Crash Course

Land use in cities in LEDCs

In the past 50 years, urbanization in LEDCs has been very rapid. Rural-urban migration has been very high with millions of people moving to cities in search of a better quality of life (see push-pull factors p.49). Cities in LEDCs have developed differently to those in MEDCs:

- as with MEDCs, the CBD is in the middle
- the pattern is similar to Hoyt's model for MEDCs, with rings and sectors
- industrial sectors follow transport routes
- the inner ring is divided into two types of housing – older, good-quality housing with good service provision and newer, high-class expensive properties with full services and amenities
- the next zone contains a variety of housing, including some former squatter areas which have been improved
- the outer zone contains shanty areas, where new migrants settle: poor conditions with few amenities.

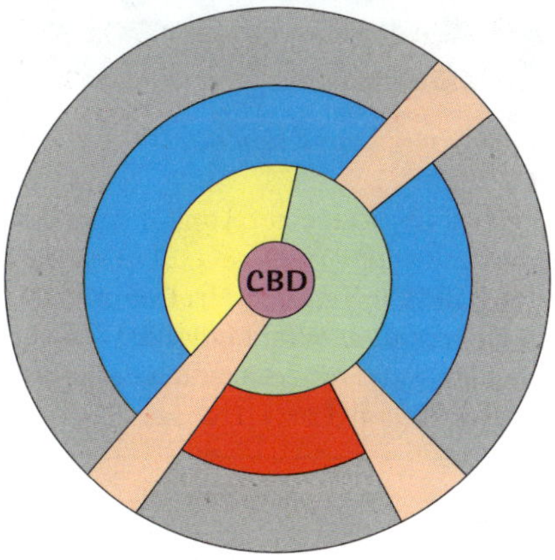

Key
- Modern, high-cost private housing
- Medium-quality housing
- Mainly older, low-cost, poorer-quality housing
- Low-cost government-funded housing improvement schemes
- Industries along transport routes
- Shanty towns

Shanty towns / squatter settlements

The outer zone on the land use model for cities in LEDCs shows an area of shanty towns. These squatter settlements are typically found on the edges of some of the fastest-growing cites in the world. These LEDC cities are growing at a far quicker rate than most cities in MEDCs have done. The outer zone is where many migrants settle when they first move to the city, usually from rural areas. They build shelters / houses out of any available material. For many people living in these areas:

- living conditions are very poor and crowded
- there are few basic amenities – often no water, sewerage system or electricity
- illness and disease are common
- health and education provision is very poor
- waste and rubbish is dumped in the streets
- many people do not have jobs – poverty is widespread
- there may be high rates of crime.

A favela on the outskirts of Salvador, Brazil

Shanty town areas are given different names in different countries e.g. 'favelas' in Brazil, 'barrios' in Mexico and 'bustees' in India. Many have existed for many years and a number have been gradually improved by:

- self-help schemes to make housing more permanent e.g. replacing wood and scrap materials with bricks and mortar, building low-cost housing units
- providing basic services e.g. electricity, water and a sewerage system
- the establishment of health clinics, schools and transport facilities
- setting up systems for loans to start small businesses e.g. shops.

However, the problem remains as there are too many people and not enough money to provide for even the basic needs of the millions who have migrated to these areas.

CASE STUDIES

EXAMPLES

Commonly used case studies:

MEDCs:
UK: Local study; Leicester; London (e.g. Docklands); Glasgow; Lakeside (Thurrock); Blue Water (Dartford); Meadowhall (Sheffield); MetroCentre (Gateshead); Merry Hill (Dudley, West Midlands)
Los Angeles, San Francisco and New York (USA); Tokyo and Osaka (Japan); Paris (France)

LEDCs:
Brasilia, São Paulo and Rio de Janeiro (Brazil); Calcutta (India); Dhaka (Bangladesh); Cairo (Egypt); Mexico City (Mexico); Nairobi (Kenya)

Sample case study 1: Rio de Janeiro - a city in an LEDC

Location:

The south east (Atlantic) coast of Brazil

Description:

- over ten million people live in Rio
- it was once the capital of Brazil
- good building land is limited, with the ocean on one side and the mountains on the other
- growth has been very rapid in recent years - the population has doubled since 1991
- traffic congestion and air pollution are high
- a fifth of the population (two million people) live in over 500 separate favelas (shanty towns) around the city
- many newcomers are migrants from rural areas in search of work and a better quality of life
- the worst favelas are built from scrap materials, have few amenities, are a long way from work in the city centre and suffer from high crime rates / drug dealing
- some of the oldest favelas e.g. Rocinha, have been improved, mainly through self-help schemes: housing has been rebuilt with bricks; there is water, electricity and sewage disposal; schools and clinics have been built; there is a bus service to the CBD; there is less crime and a better quality of life
- people still migrate to Rio, but the population increase has slowed down as others move out
- a new town, Barra, has been built 20km away from Rio - attracting people to move
- Barra has all the advantages of a new town - modern housing and shopping facilities, entertainment and leisure facilities, easy access, a safer environment and better quality of life

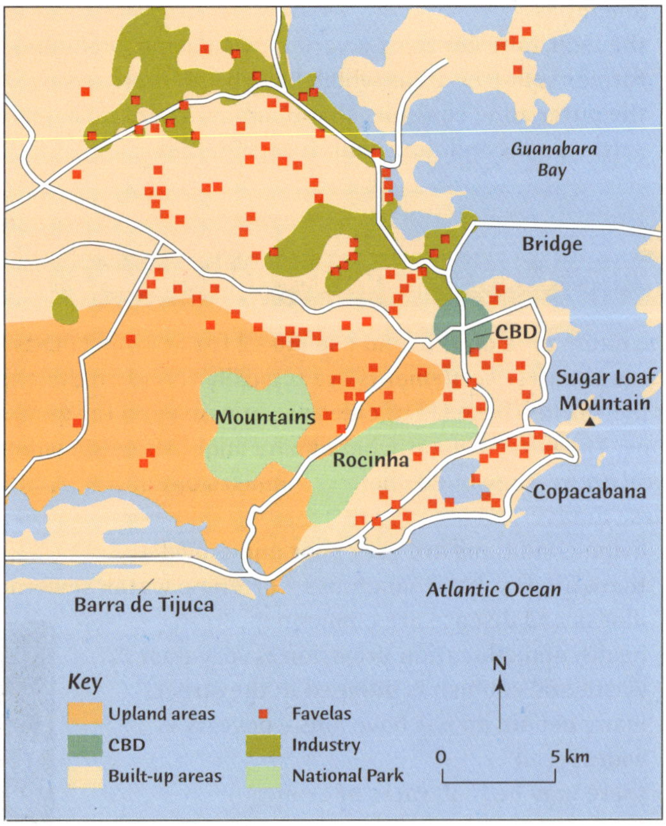

Rio de Janeiro

Rio is a typical example of a city in an LEDC. Its population has grown rapidly, putting pressure on land and resources. Many of the old favelas have been improved, but millions still live in very poor conditions. New developments like the town of Barra have caused many to move away - but these are usually people with the money to do so and the skills to find jobs there.

CASE STUDIES

Sample case study 2: The MetroCentre, Gateshead - shopping services

Location:
South west of Newcastle on the A1 western by-pass; just south of the River Tyne

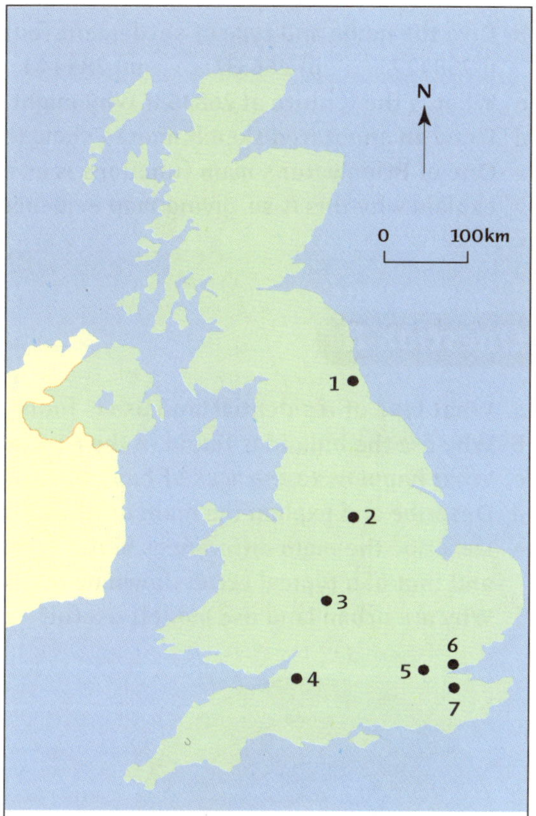

Description:

- built on the site of a derelict power station: cheap flat land
- plenty of space to build and expand
- one of the largest shopping centres in Europe
- within an Enterprise Zone (financial and planning incentives)
- the idea of Sir John Hall, local businessman
- close to main roads e.g. A1
- free parking space for over 12 000 cars
- good bus and rail links - own railway station
- large catchment area - nearby Newcastle, Gateshead and the surrounding urban area
- a vast range of shops (over 300)
- attractive, light and under cover
- places to eat and drink; entertainment (for adults and children); and child-care facilities
- large luxury hotel on site

Regional shopping centres in the UK today

There has been much discussion about the impact of both regional centres and out-of-town shops and supermarkets since they began to increase in number during the 1980s. Although they offer good value, with competitive prices and a great range of goods, they have had a great impact on smaller shops and on city centres where shops have closed down and sites become derelict. Usually built on the outskirts of urban areas, they have sometimes created conflict in once quiet areas with heavy lorries delivering goods and thousands of shoppers arriving by car seven days a week. Some city centres are now beginning to revive - with pedestrianization and covered precincts, shoppers are being attracted back. Planners may find it more difficult in future to get the necessary permission to build out of town.

EXAM PRACTICE

Question A

Study the OS map on page 92.

a Name and give six-figure grid references for two leisure facilities in or near Cannington Village (260395).

b Give the name and type of settlement found at:
i) 258422 ii) 266337 iii) 288444.

c What is the feature at 268363? Why might it be important to Bridgwater?

d Draw an annotated sketch map to show the site of Bridgwater.

e One of Bridgwater's main functions is as a market town. Write a paragraph to explain why this is so, giving map evidence to support your answer.

Question B

a What type of residential land use is found at A, B and C in models 1 and 2?

b Why are the buildings found in the CBD often the tallest in the town or city?

c What happens to the ages of buildings as you move out from the CBD?

d Describe and explain the main differences between model 1 and model 2.

e Describe the main differences between the structure of cities in MEDCs shown by models 1 and 2, and that of a typical LEDC shown by model 3.

f Why are urban land use models useful?

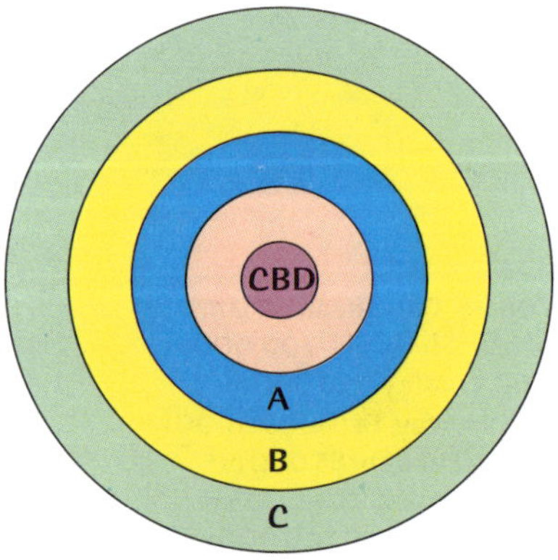

1 Concentric model (Burgess)

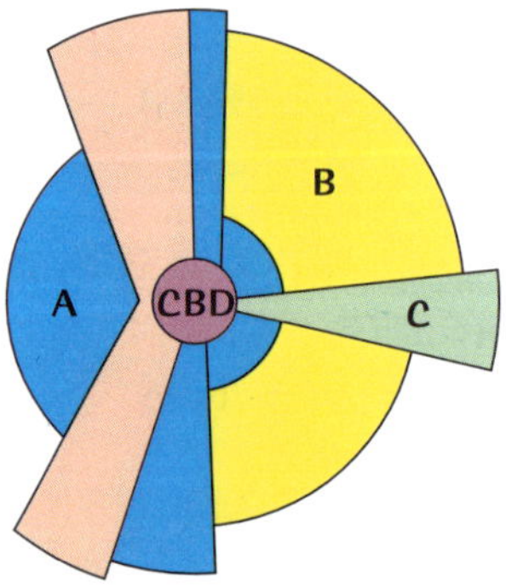

2 Sector model (Hoyt)

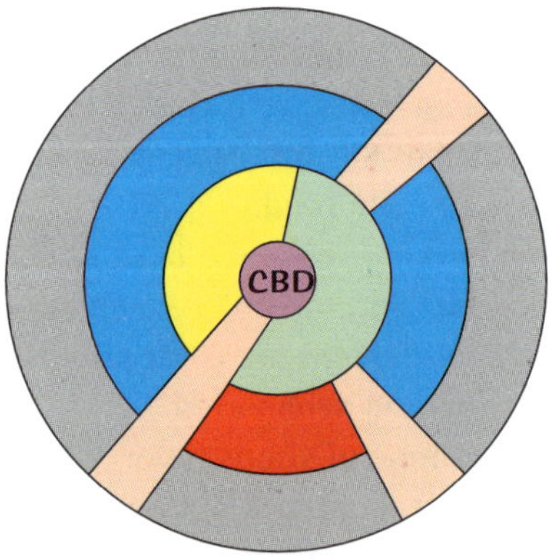

3 LEDC urban model

Question C

Part (1)
Study the photograph of the shanty town. Describe some of the problems of living in this sort of environment.

Part (2)
For a city in an LEDC you have studied, where attempts have been made to improve housing:

a Name the city and describe its location.

b Describe what the area was like before improvements were made.

c Describe and explain what improvements have been made and by whom.

A shanty town on the edge of Manilla in the Philippines

Revision Unit 9

Economic activity

The Check up

1 Industry can be divided into three types – __________, secondary and __________.

2 Study the two divided bar graphs, then complete the statement.

Graph __________ is typical of the employment structure of a __________ country because it has a high proportion of workers in the __________ sector.

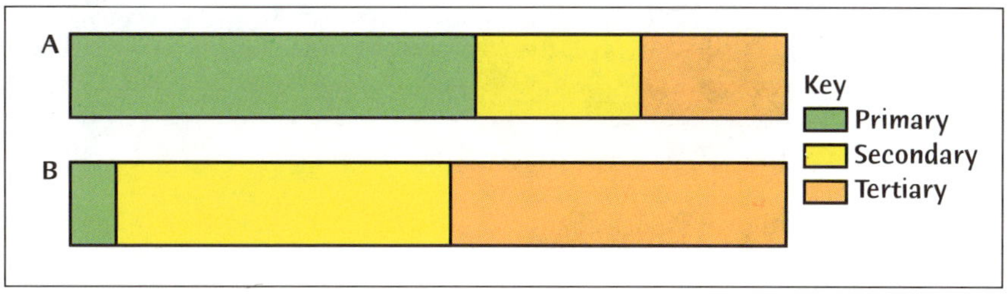

3 Three of the factors influencing the location of industry are __________ (money), __________ (customers) and __________ (workforce).

4 Industries can be seen as systems made up from inputs, __________ and outputs.

5 __________ is the world's most important industry, covering 40% of the land surface and employing over 45% of the world's population.

6 Growing crops is called __________ farming: rearing __________ is called pastoral farming.

7 __________ herders move their animals from place to place in search of food.

8 Artificially watering land to grow crops is called __________.

9 __________ agriculture is the name given to growing large areas of single crops e.g. coffee, rubber, in the Tropics.

10 In the 1960s the __________ Revolution began, trying to improve agricultural production in __________ countries.

11 In the EU, the __________ guarantees production levels and prices, sometimes leading to __________.

12 __________ industry involves the use of bulky __________ materials e.g. iron and steel making.

13 The car industry is a good example of a secondary industry using the __________ line process to produce finished goods.

14 __________ are large companies which operate in many different countries.

15 The increase in industry organized on a world scale is known as __________.

16 Industry is a major cause of environmental pollution – especially the production of greenhouse gases which contribute to __________ warming and __________ rain.

17 __________ industries include biotechnology, pharmaceuticals and aerospace industries.

18 Much of the UK's older industries were located on __________.

19 __________ locations are common today, making __________ and exporting more convenient.

20 Attractive out-of town sites for industry are known as __________ sites.

Mark yourself *(see answers on page 100)*

- More than 12? Do the Exam Practice – or go on to another unit.
- Less than 12? Do the Crash Course on the next page.

/ 20

The Crash Course

Key facts to learn: Types of industry

The jobs people do can be divided into three main groups:

- **Primary**: collecting or producing raw materials e.g. farming, fishing, forestry, and mining
- **Secondary**: construction, manufacturing or processing raw materials to produce goods e.g. cars, electricity
- **Tertiary**: providing a service e.g. health, transport, education, financial, retailing

A fourth group, **quaternary**, is sometimes used. This refers to high-level service jobs e.g. research and information services.

Employment structure

The proportion of people working in each type of industry is called employment structure. Figures are shown as percentages, drawn as divided bar graphs or pie charts. These graphs and figures can be used to show the level of development in different countries. LEDCs tend to have a high proportion in primary industry (usually farming), whereas in MEDCs the highest proportion is usually found in tertiary industry. As countries start to develop their economies, secondary industry increases as primary industry decreases. Official figures hide one aspect of employment – the difference between formal and informal work. Statistics usually refer to formal paid work (wages); many people (especially in LEDCs) are in informal work.

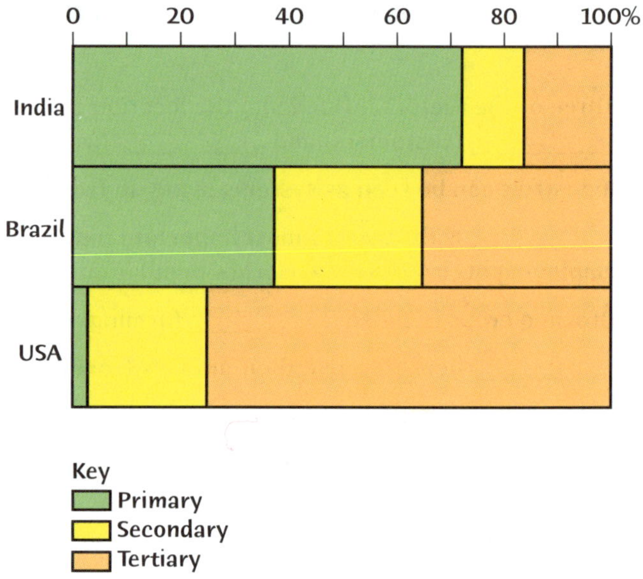

Industrial location

An industry or factory will look for the most profitable place to locate. This will involve a range of factors – the type of industry will determine which factors are the most important e.g. an iron and steel works will usually locate near to the bulky raw materials needed (iron ore, coal) as they are expensive to transport. If these raw materials are imported, a coastal location may be chosen.

Many industries can be seen as a system of **inputs** (raw materials, energy, capital, labour), **processes** (manufacturing the final product from the raw materials) and **outputs** (the final product).

THE CRASH COURSE

Primary industry: agriculture (farming)

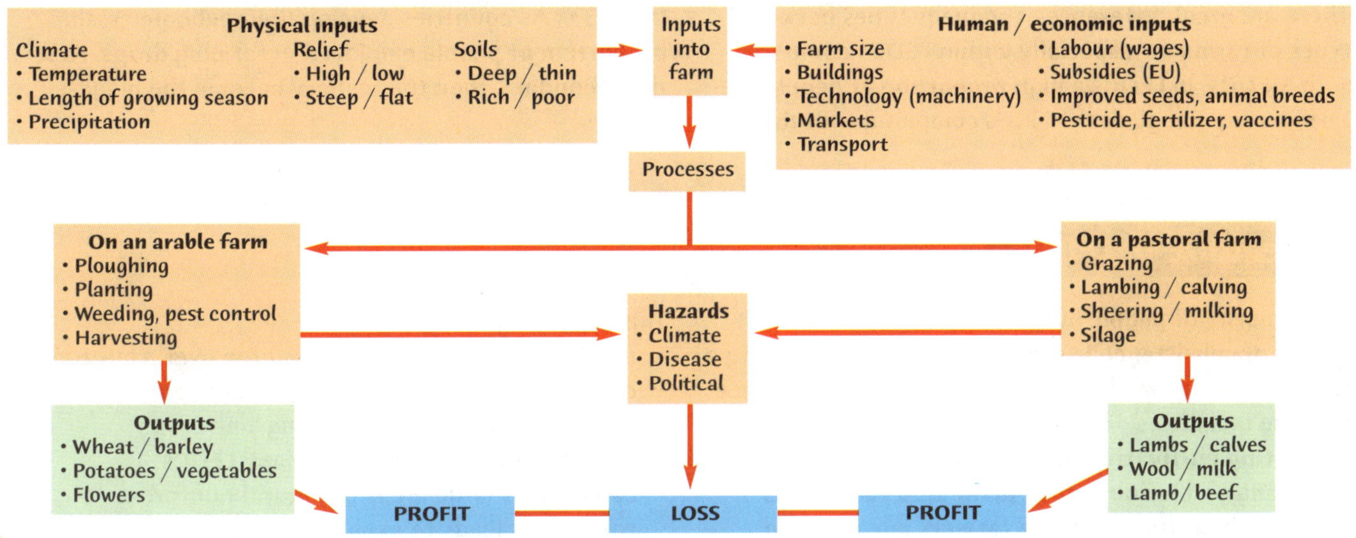

KEY FACTS TO LEARN ABOUT FARMING

- Farming is the world's most important industry.
- It supplies us with the food we need to survive.
- 40% of the world's land surface is used for farming.
- Farming includes rearing animals and growing crops.
- Over 45% of the world's population is employed in farming.
- Like other industries, farming works as a system of inputs, processes and outputs.

Types of agriculture

- Arable: growing crops
- Pastoral: rearing animals

- Subsistence: producing just enough food for your own / your family's consumption
- Commercial: producing enough to sell to make a profit

- Nomadic: moving from place to place, especially animal herds e.g. cattle and goats
- Sedentary: farming in one place

- Intensive: high inputs into a small area producing high yields / outputs
- Extensive: low inputs over a large area producing low yields / outputs

Farming can be a mixture of these types – rearing cattle may be a commercial process e.g. in Texas, USA. It could also be subsistence and nomadic e.g. Masai cattle herders in Kenya.

Other farming terms

- Dairying: rearing cattle for dairy products e.g. milk, butter, cheese
- Irrigation: artificially watering the land to allow farming to take place
- Cereals: grain crops e.g. wheat, barley, oats
- Mixed: growing crops and rearing animals
- Market gardening: intensive farming of fruit, vegetables, salad crops and flowers

THE CRASH COURSE

Primary industry: agriculture (farming)

There are great differences and inequalities between types of farming - especially within LEDCs compared with MEDCs. In LEDCs a high proportion of people are farmers e.g. Bangladesh = 70% compared with the UK = 3%. As countries develop their economies, the proportion of people employed in farming drops. First the secondary, then the tertiary become the biggest sectors.

Agriculture in LEDCs

The majority of people living in LEDCs are farmers - mainly at subsistence level growing crops and rearing a few animals on small pieces of land. Other subsistence farming includes:

- **shifting cultivation**: farmers clear an area by cutting and burning, then plant seeds and crops. After about five years the farmers move on and clear another area as the soil loses its fertility.
- **pastoral nomadism**: herding livestock from place to place in search of pasture / food.

In most basic types of farming, the purpose is to provide enough food for the farmers and their families, perhaps earning some money by selling any surplus. In Tropical regions, **plantation agriculture** is common.

This type of farming:

- involves the cultivation of one crop over a large area e.g. bananas, cacao, tea, rubber, sugar cane
- is called monoculture (meaning 'one crop')
- provides jobs and money to many LEDCs
- can destroy large areas of natural rainforest
- sometimes leads to exploitation of workers
- can take profit out of the country if the plantation is owned by a foreign company (e.g. a TNC).

Rapid population growth has increased the pressure to produce more food, often under very difficult conditions e.g. lack of money for investment, poor soils, flooding and drought.

The Green Revolution

The Green Revolution:

- introduced new farming methods to LEDCs in the 1960s
- attempted to increase food production via **intensive** farming
- moved from labour intensive to capital intensive
- developed the use of high yielding crops
- used machinery to plough fields
- set up irrigation schemes and projects.

Yields did increase, but there were problems - a range of (expensive) fertilizers and pesticides had to be used which posed a possible threat to health; high levels of irrigation were needed; jobs were lost to mechanization. Today, more appropriate methods of improving farming are being developed e.g. using **intermediate technology** which is **sustainable**.

Agriculture in MEDCs

Although there are many farmers in MEDCs who do not make large profits e.g. hill sheep farmers, most agriculture in MEDCs is:

- commercial (often called agribusiness) - concerned with high levels of production and profit
- usually capital intensive
- highly mechanized (where possible) - not labour intensive
- often supported by subsidies, quotas and trade tariffs. In the EU, the **Common Agricultural Policy** (CAP) guarantees production and prices (for producers and consumers). Unfortunately this has led to overproduction and land **set aside** (farmers paid not to farm land or produce crops).

The Crash Course

Secondary industry

Secondary industry involves making or **manufacturing** a finished product from raw materials. It also includes the **construction** of buildings, roads, bridges etc. Types of secondary industry include:

- **heavy industries** which use large quantities of often bulky (heavy) raw materials e.g. shipbuilding, iron and steel
- **light industries** which use smaller amounts / less raw materials / small components to produce e.g. cameras, TVs
- **high-tech industries** which produce high value, expensive products made using the latest technology e.g. computers, satellites.

Some of the most common secondary industries are shown in the table:

Industry	Main raw materials	Products	
Iron and steel	Iron ore, coal, limestone	Sheet steel, steel girders	
Cars / vehicles	Sheet steel, plastics, rubber, glass and thousands of parts and components	Cars, vans, lorries	
Shipbuilding	Sheet steel, glass, plastics and thousands of parts and components	Ships, boats	
Electrical goods	Metals, plastics and many small parts and components	TVs, computers, washing machines, cameras, microwaves etc.	
Food processing	A range of vegetables, fruits and animal products; sugar; salt; vegetable and animal oils and fats	Processed food / drink e.g. chocolate, breakfast cereals, fish fingers etc.	

Tip: Remember that most factories also use vast quantities of water (for manufacturing processes, cooling and cleaning) and power (usually electricity).

The car industry

The car industry is a good example of a secondary industry and also of how this type of industry can be seen as a system of inputs, processes and outputs – a good way to describe how most manufacturing industries / factories work.

Processes:
Assembly
Painting/spraying

Inputs:
Body parts
Engine
Tyres
Electrics
Glass (windows)

Outputs:
Car

In the car industry both raw materials and thousands of **components** or parts (usually made elsewhere) are put together or **assembled** (on assembly lines) by people and / or machines (robots) to produce the finished product – a car.

The Crash Course

Transnational Companies (TNCs)

Transnational (or multinational) companies are very large companies which operate in several different countries. The headquarters of the company are usually in an MEDC, with factories in both MEDCs and LEDCs. The world's biggest TNCs have an annual income higher than many countries and exert great economic influence and power. They are able to invest in research and development and modern and efficient production methods yet offer competitive prices.

Globalization:

- companies which trade across the world
- industry organized on a global or world scale
- originally began as TNCs in MEDCs; started to operate on a wider scale
- LEDCs are now starting to develop their industries on this scale.

Advantages

TNCs may:

- provide training, new skills and jobs
- develop the infrastructure e.g. roads, services
- employ workers who earn and spend money, pay taxes – helping their country's economy
- boost other industries in the host country (multiplier effect).

Disadvantages

TNCs may:

- take money (profits) out of the country
- employ local people in the lower-paid jobs, bringing in their own more highly-paid managers
- make goods mainly for the export, not local market
- close overseas factories first in times of recession.

When building a factory in another country, a TNC can benefit in a number of ways:

- lower costs e.g. cheaper land and wages;
- develop new (nearby) markets e.g. building a factory in Spain can open up the European market for a Japanese or American company;
- avoid import duties or tariffs, lowering the cost of their products.

Environmental issues

Industry is essential to produce the energy and goods we need. However, it also has a great impact on the environment – something which people are becoming more aware of today. Locally, discharges and waste from industry can cause pollution to streams, rivers and the atmosphere. Noise and visual pollution may also occur. Globally, a number of environmental problems result from industrial processes:

- **acid rain**: emissions from power stations pollute the atmosphere, causing widespread damage to e.g. forests and lakes when they mix with rain
- **ozone depletion** or thinning: CFCs are damaging the ozone layer, increasing the problem of damage from UV light
- **global warming**: greenhouse gases from industry are thought to be causing global temperatures to rise, leading to a number of potential problems e.g. sea level rises.

The effects of acid rain on a coniferous forest in Scandinavia

EIAs and Agenda 21

Today, before permission is given for a factory to be built, local authorities or governments usually insist that an **Environment Impact Assessment (EIA)** is carried out before they will consider granting planning permission.

Agenda 21 originated at the Rio Earth Summit of 1992. It identifies what needs to be done to encourage sustainable development in the twenty-first century, including alternative energy supplies, preserving ecosystems and protecting the environment.

The Crash Course

High-tech industries

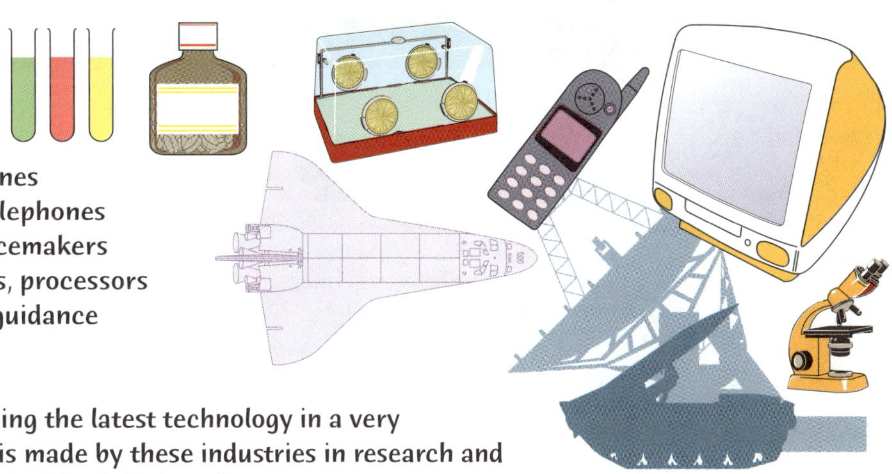

High-tech industries include:

- biotechnology
- pharmaceuticals – drugs and medicines
- telecommunications equipment – telephones
- medical equipment – incubators, pacemakers
- information technology – computers, processors
- defence equipment – radar, missile guidance
- aerospace – satellites, spacecraft.

These are high value products made using the latest technology in a very competitive market. Much investment is made by these industries in research and development, with highly qualified and highly skilled workers.

Key facts to learn on high-tech industries:

- use small quantities of raw materials
- need a skilled workforce
- are located close to motorway networks for accessibility
- are often found grouped in **Business** or **Science Parks**
- work closely with academic institutions like colleges and universities, sharing expertise and facilities
- are sometimes called **footloose** industries as they can thrive in a wide variety of locations.

Changing locations

When industry developed in the UK, during the Industrial Revolution which spanned the nineteenth and early twentieth centuries, it began and then concentrated on or near the most important raw material of the time – coal. (See map 1.) Today, with very little coal being mined in the UK and raw materials largely imported from abroad, many of the old heavy industrial areas have declined.

Map 2 shows where today's industry is concentrated: on coastal sites for easy import and export and on greenfield sites outside urban centres. Greenfield sites are attractive, often landscaped, sites where workers enjoy modern facilities and conditions in 'smart' computer-controlled buildings. Most high-tech industries, Science and Business Parks are found in these locations. Newer shopping centres, supermarkets and superstores have similarly moved to out-of-town greenfield sites.

Map 1

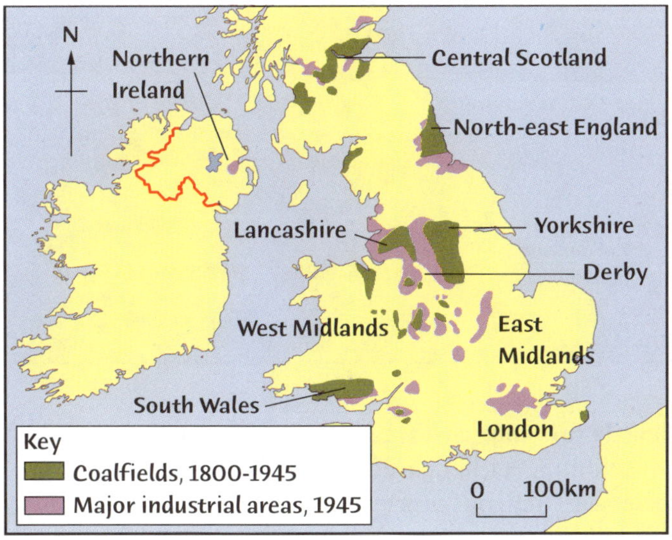

Map 2

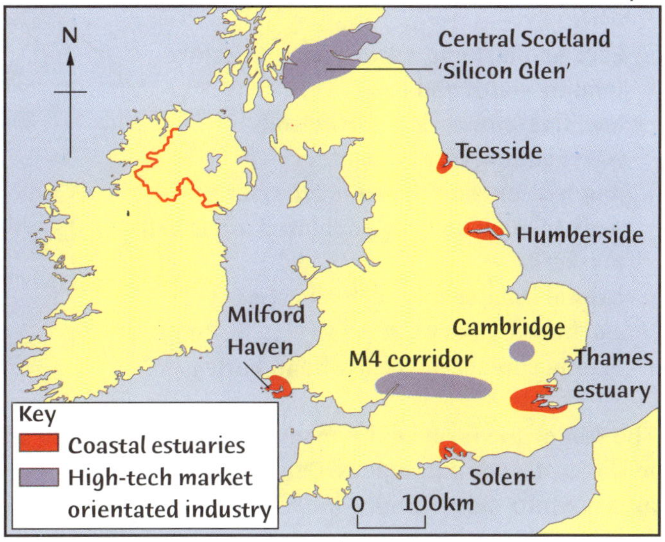

The Crash Course

Examples

Commonly used case studies:

Farming:

UK:
local studies; crofting (Scottish Highlands); arable (East Anglia); hill / sheep farming (Lake District)

MEDCs:
dairying and mixed (Denmark); subsistence and fruits (Mezzogiorno, Italy); dairying and horticulture (Netherlands); Japan; Rhine-Ruhr (Germany)

LEDCs:
subsistence (India and Bangladesh); shifting cultivation and plantation (Brazil); rubber (Malaysia); nomadic herding (Sahel)

Industry:

UK:
South Wales; London Docklands; M4 corridor; Silicon Glen (Scotland); Science Parks (Cambridge and Newport); quarrying (National Parks); iron and steel (Teesside); pottery (Stoke)

MEDCs:
Hiroshima, Toyota City and Tokyo (Japan); Detroit and California (USA); Lille (France); port - Rotterdam (Netherlands)

LEDCs:
Pacific Rim / South Korea - Malaysia; Nairobi / Mombasa (Kenya); mining (Amazonia, Brazil); São Paulo state industry (Brazil); textiles (India)

Case Study 1: Farming in the UK

Location:

The map shows the distribution of the main types of farming in the UK.

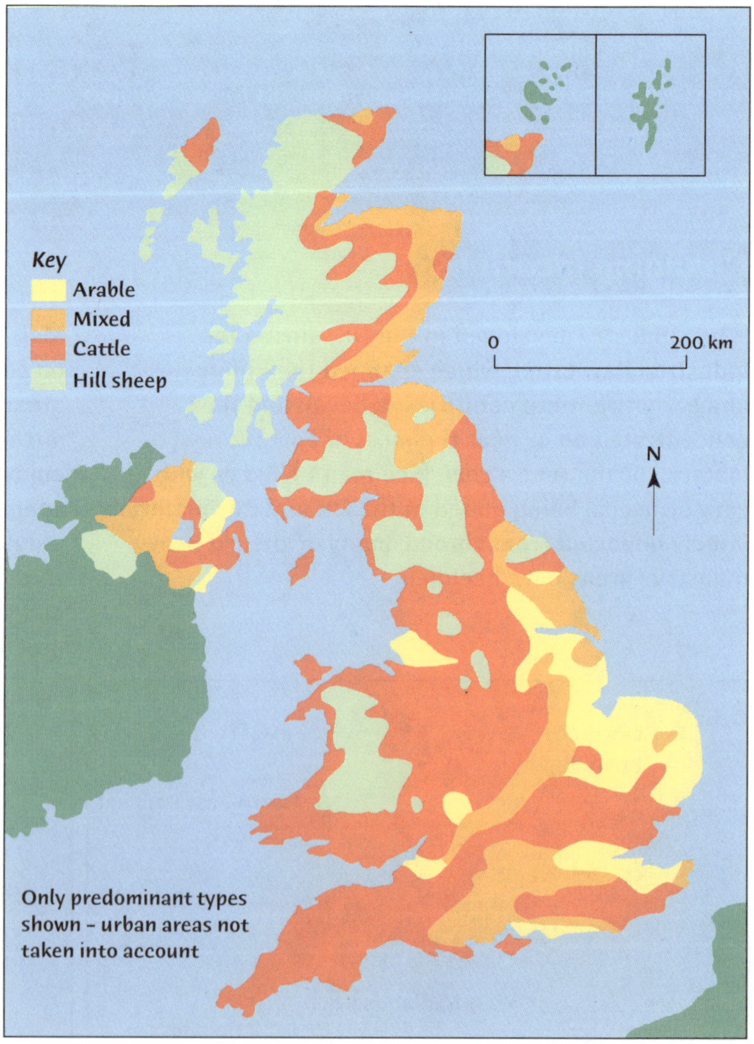

Description:

There are quite clear patterns of distribution for the main farming types, with physical factors such as climate and soils and human factors such as markets and capital being important.

Hill farming:

- Scottish Highlands, Pennines, Lake District, Welsh mountains, Dartmoor and Exmoor (mainly north and west)
- steep slopes = poor, thin soils
- mainly cool summers, cold winters
- snow and heavy rain
- isolated
- mainly sheep farming

Arable:

- East Anglia, Kent, Essex, Avon, Cheshire (mainly south and east)
- low, flat land = rich, fertile soils
- warm summers, cold winters
- much drier than the west; summer rain
- near to main densely populated areas and markets
- capital intensive - much investment e.g. in machinery
- mainly cereal crops e.g. wheat, barley

The rest of the farm land is mainly pastoral - dairying and cattle farming, with areas of mixed farming mainly in the east / central regions. Other types of farming e.g. horticulture, market gardening are found near built-up areas. Whilst most farming in the UK is on a commercial scale, with high outputs yet a small workforce (mechanization), many farmers struggle to make a living e.g. hill sheep farmers.

Case Studies

Case Study 2: The car industry - Toyota, a TNC (Trans National Company)

Location:

In Japan, all Toyota's factories, except one, are found in Toyota City, west of Tokyo. This includes five assembly and two engine plants plus many component factories and the company's head office.

Toyota time line:

1937 Toyota Motor Company founded which produced Japan' s first passenger car

1947 Toyota's first small car produced via mass production

1950 Company split into two parts: manufacturing, and sales/marketing

1960's New production systems - 'just in time', Kanban and TQC (total quality control) introduced

1970's Toyota's production systems named TPS (total production system), empowering workers and improving factory production.

1991 Toyota became the world's leading motor car producer

Description:

A good example of a TNC (Trans National Company), Toyota has a global workforce of over 100 000 and 35 factories in more than 20 countries. It became the world's leading car manufacturer in 1991. It has two factories in the UK - an assembly plant in Burnaston, Derbyshire and an engine plant on Deeside, North Wales. Burnaston covers almost 600 acres, employs over 3000 people and produces over 170 000 cars per year. Deeside is a smaller factory covering just over 100 acres, employing 200 people and producing 100 000 car engines per year. About three quarters of the cars produced are sold to the rest of Europe. Toyota chose these sites because:

- the large UK and European markets are close by
- a skilled and flexible workforce was available
- government support was given at both local and national level, though no financial support
- there is a good communications network
- a large piece of flat land was available
- there is a strong local tradition of car manufacturing and component supply.

Toyota's Burnaston plant, Derbyshire

EXAM PRACTICE

Question A

Study diagram (a) showing the farming year in Bangladesh.

a For each of the three different rice crops, Boro, Aus and Aman:
  i) what is the growing season?
  ii) describe the climatic conditions during the growing seasons.

b i) Farms are small, but yields are high. Name this type of farming.
  ii) How has the Green Revolution helped areas like this?

c Many Bangladesh farmers are subsistence farmers. Describe what this means and the sort of crops and animals which might be found.

d What are the problems of subsistence farming?

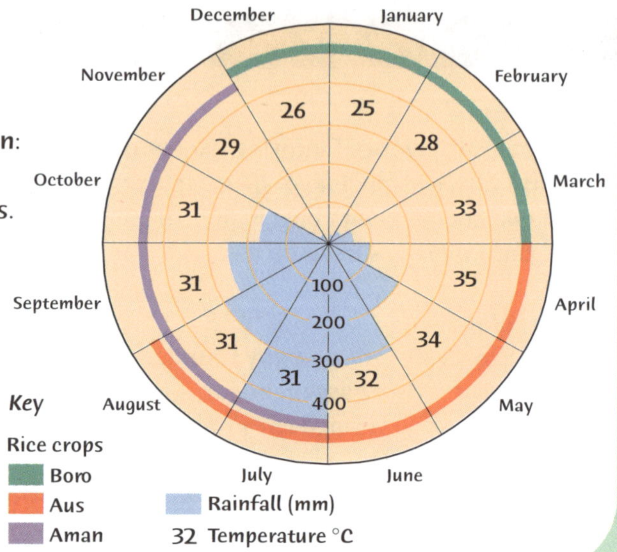

Diagram (a)

Question B

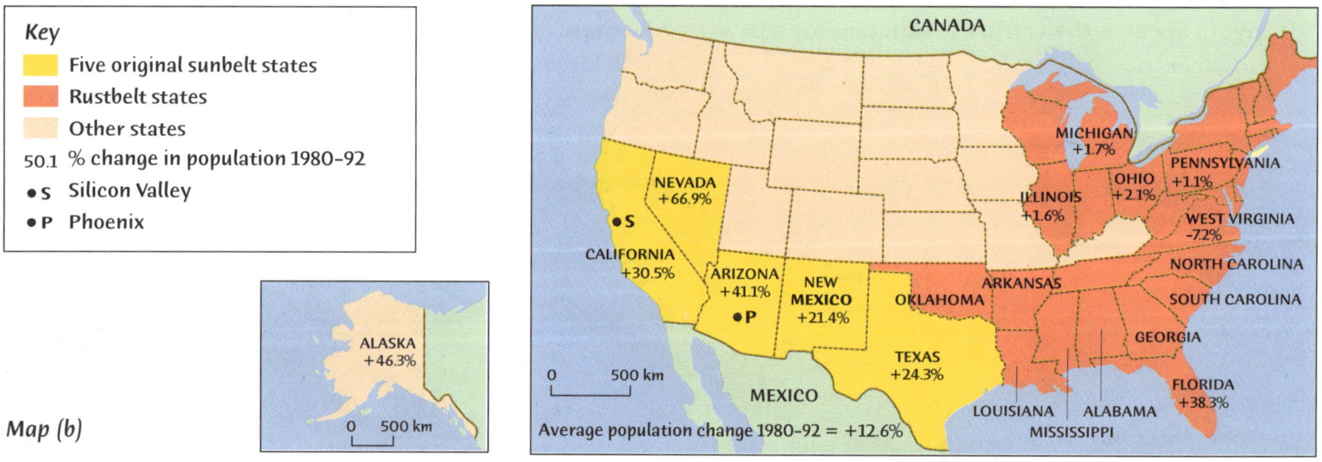

Map (b)

Study map (b).

a Describe the location of:
  i) the rustbelt states
  ii) the sunbelt states.

b The USA's industry developed first in the rustbelt states, based on heavy industries. Today, new high-tech industry is located in the sunbelt states.
  i) What is meant by heavy industry and high-tech industry? Give examples.
  ii) Describe and account for the population changes between the rustbelt and sunbelt states.
  iii) What problems might these changes bring?

c For a named area you have studied (not the USA), describe how the location of industry has changed. Give reasons for the changes.

Question C: case study question

For a named TNC you have studied:

a Name the TNC and its country of origin.

b What does it produce?

c For a chosen country / location where the TNC has set up (outside the home country):
  i) explain why it located there
  ii) describe the benefits and disadvantages it has brought to the company and the country.

Revision Unit 10

Resources and energy

The Check up

1 Resources are either __________ or non-renewable.

2 Non-renewable resources are finite – they can only be used __________ and cannot be replaced.

3 __________ resources can, with careful management, be used again and again – they are __________.

4 Some of the most important resources are __________ resources, used to power machinery, equipment and transport.

5 Coal, oil and gas are all __________ fuels, formed from the remains of plants and animals.

6 Study the two divided bar graphs (a), then complete the statement. Only __________ of the world's population live in __________, yet they use __________ of the world's energy resources.

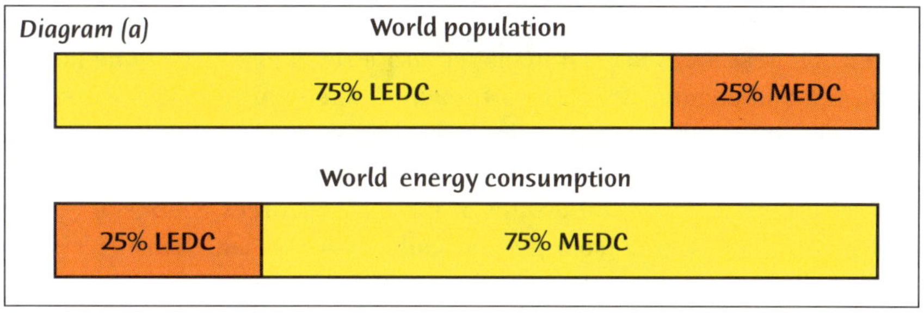

7 Energy consumption is a good __________ of development.

8 Burning coal, oil and gas creates __________ gases which contribute to __________ warming and __________ rain.

9 __________ and __________ are often found together underground in layers of sedimentary rock.

10 Heat produced by __________ in a chain reaction creates nuclear power.

11 Although nuclear power is much cleaner than other non-renewable sources, people are concerned about the dangers of __________ and the disposal of nuclear __________.

12 Millions of people in LEDCs rely on __________ for heating and cooking.

13 __________ energy sources provide just a small proportion of world energy production at present.

14 Look at diagram (b). Name the features at A, B and C.

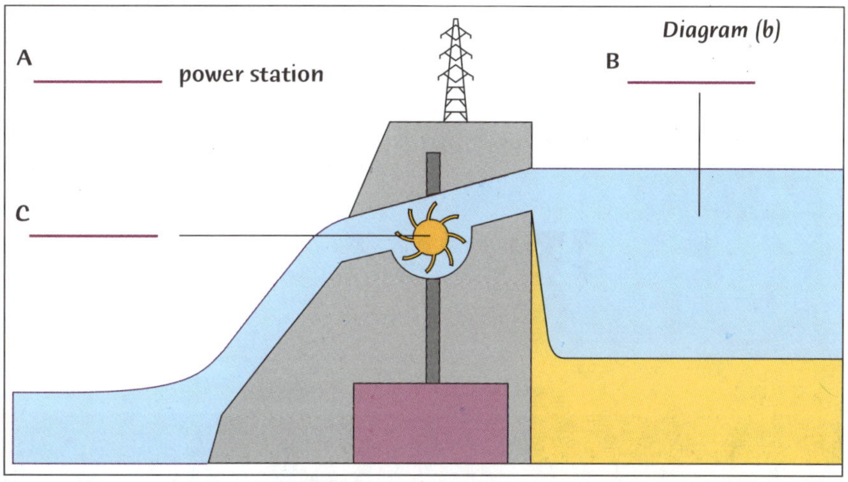

15 Energy can be produced by using the wind to turn __________ to produce electricity.

16 __________ power uses energy from the sun via photo cells and panels.

17 Heat from volcanic activity is used to produce __________ energy.

18 Gases given off by __________ and __________ waste can be used to produce energy (biogas).

19 Producing and using the __________ we need for homes and industry often causes great __________ to the environment.

20 Spills from __________ tankers can pollute large areas of coastline.

Mark yourself *(see answers on page 101)*

- More than 12? Do the Exam Practice – or go on to another unit.
- Less than 12? Do the Crash Course on the next page.

/ 20

THE CRASH COURSE

Resources and energy

Natural resources are things in our environment which we use, and are useful to us. Resources can be divided into two groups:

Non-renewable

Non-renewable resources are **finite** – once they have been used up, they cannot be replaced. They have usually been formed over very long periods of time, under specific conditions e.g. coal, oil, gold and diamonds.

Renewable

Renewable resources **can be used again and again** because they are continuous or replaced fairly quickly e.g. solar and wind power, trees and water. However, if renewable resources are not managed properly, they can become damaged or depleted and become difficult to sustain or keep going. Managing resources carefully so that they are **sustainable** is important.

One of the most important groups of natural resources is **energy**. All the Earth's energy resources come from the sun and provide us with warmth and light, and work machinery, equipment and transport.

World energy production

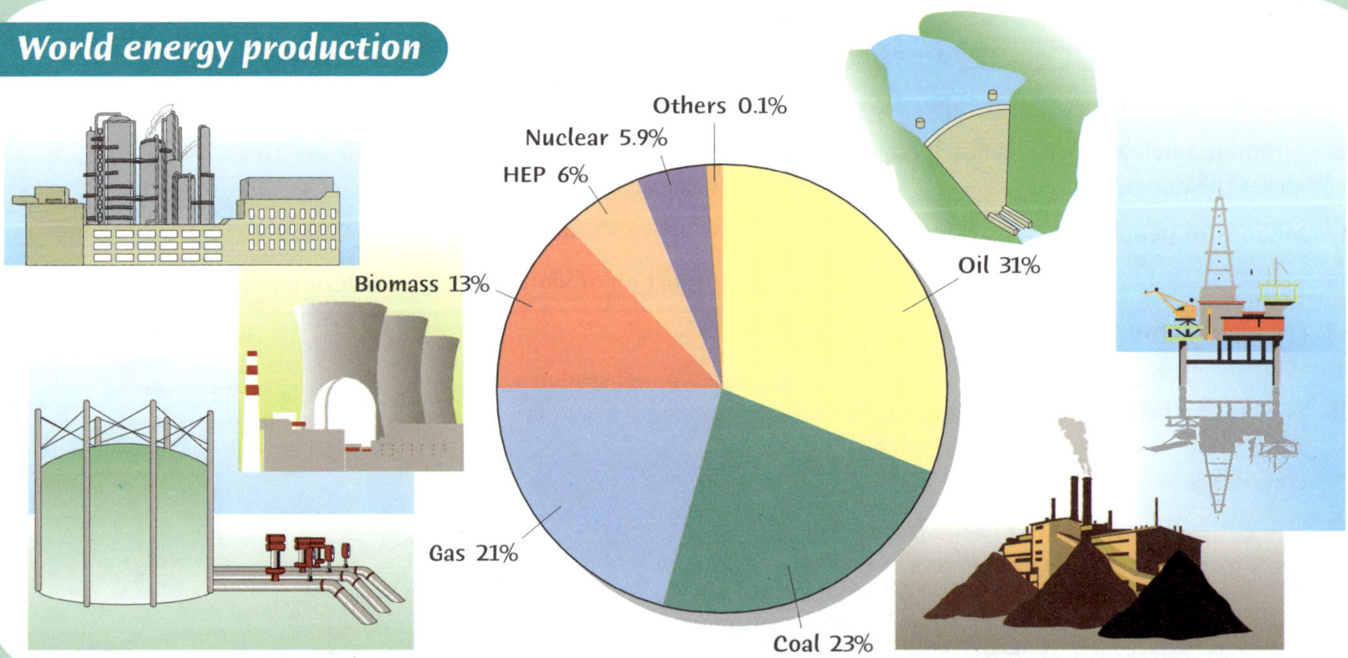

KEY FACTS TO LEARN ABOUT WORLD ENERGY CONSUMPTION:

- 75% of the world's energy is used by MEDCs; yet only 25% of the world's population live in MEDCs.
- most of the world's energy resources are found in MEDCs. Energy was one of the major factors enabling these countries to industrialize and develop their economies
- few LEDCs have substantial energy resources
- many non-renewable resources in MEDCs are running out
- the amount of energy used by a country is a good indicator of its level of development
- LEDCs rely heavily on trees for fuel. Although these are renewable resources, the rate at which they are being cut and used is unsustainable.

THE CRASH COURSE

Non-renewable energy

- 75% of the world's energy comes from oil, coal and gas which are fossil fuels made from plant and animal remains. They are non-renewable.
- Burning fossil fuels releases large quantities of the 'greenhouse gases' into the atmosphere, leading to environmental problems such as global warming and acid rain.
- Nuclear power, once seen as a cleaner alternative, has attracted fierce public opposition because of fears over safety following radiation leaks at Chernobyl and the difficult problem of disposing of waste safely.
- The cutting and burning of fuelwood, mainly in LEDCs, has led to widespread deforestation and other environmental problems.

The cooling towers of a typical thermal power station

Summary of non-renewable energy

Type	Description and uses	Main producers	Advantages	Disadvantages
Coal	Formed from decaying plants in layers underground. Largely used to make electricity today.	Western Europe, USA, China, former USSR, India, South Africa	High levels of reserves; new mines are highly mechanized	Pollution – produces high levels of CO_2, a greenhouse gas; underground mining can be dangerous
Oil	Formed from the remains of animals, often trapped in layers of sedimentary rock underground. Used for fuels, plastics, fertilizers, medicines and electricity production.	USA, China, Middle East, Mexico, UK, Canada, Norway, former USSR, Venezuela	Many uses; fairly easy to transport (tankers and pipes); efficient to burn	Low reserves; pollution; danger of oil spills e.g. *Exxon Valdez*; toxic
Natural gas	Similar to oil formation, and frequently found with oil deposits. Used for heating, cooking and electricity production.	USA, China, Canada, Mexico, Venezuela, Algeria, former USSR	Cleaner than other fossil fuels, easy to transport (tankers and pipes)	Highly flammable; air pollution
Nuclear	Heavy metal – uranium – an element naturally occurring in rocks. Used in a controlled chain reaction to produce heat for electricity production.	Canada, USA, Spain, Namibia, South Africa, Australia, former USSR	Cleaner than fossil fuels; efficient; less bulky waste products	Radiation; public fears over safety (Chernobyl); disposal of waste; cost of making old reactors safe
Fuelwood	Trees from the local area, not specifically grown for fuel. Used to provide heat and for cooking (competing with other uses – building fences, houses etc.).	Mainly collected in LEDCs from natural environment e.g. sub-Saharan Africa	Often available nearby; can be collected daily; free. Can be sustainable if properly managed	Time-consuming daily collection; deforestation can lead to soil erosion and land degradation

The Crash Course

Renewable energy

- Generally much cleaner and more environmentally-friendly than non-renewable sources.
- Can be sustainable / most won't run out.
- Expensive at present.
- Should become cheaper as technology becomes available to develop sources.
- Currently accounts for a very small proportion of the world's energy production.

A wind farm with hundreds of turbines

Summary of renewable energy

Type	Description and uses	Main producers	Advantages	Disadvantages
Solar	Photovoltaic cells and solar panels use sunlight to produce electricity. Used for electricity and direct heating	USA, India	Unlimited supplies, especially in hot climates; efficient and clean	Equipment needed is expensive at present; needs sun so daytime only
Wind	Wind blows, turning blades of large turbines to generate electricity	Denmark, California, small scale in many other countries e.g. UK	Very clean, no air pollution; cheap to run - free winds!	Noisy; visual pollution; winds are not constant
Hydro-electric	Fast-flowing water turns turbines and so generates electricity. Need high volume and constant water supply (dam, reservoir and steep gradient)	Canada, USA, former USSR, Brazil, China	Very clean; reservoirs can be multi-purpose; usually in remote, sparsely-populated highlands	Large areas flooded for storage destroying habitats; silt trapped behind dam; visual pollution
Tidal	Moving tidal water (usually at river estuaries) used to power turbines and so generate electricity	France, China, Canada, former USSR	Large-scale schemes = high output; clean; can help prevent coastal flooding	Very expensive to construct; affects coastal habitats; few suitable sites
Geo-thermal	Natural heat from volcanic activity. Boreholes drilled and cold water pumped down - steam and hot water pushed back, used for direct heating and to generate electricity	Japan, New Zealand, Iceland, Hungary, former USSR	Much potential; many possible sites around the world - free heat!	Expensive, bi-product = sulphur gases; dangers from very high temperatures
Biogas / biomass	Fermenting plant and animal waste gives off gases - ethanol, methanol. Can be used for heating / to generate electricity	Brazil, Japan, India, Germany, Denmark	Widely available, especially in LEDCs; uses waste products	Waste can no longer be used as fertilizer; large schemes are expensive to set up

Environmental issues

Using natural resources to produce the energy necessary for home and industrial use has a great impact on the environment, in terms of pollution and depletion of resources (e.g. see Case Study, page 75).

CASE STUDIES

EXAMPLES

Commonly used case studies:

MEDCs:
UK local studies (various); South Wales (coal); UK (electricity); Cumbria, UK (wind); France (nuclear and tidal); USA (oil / fossil fuels); Alaska, USA (oil); Japan (nuclear); New Zealand (geothermal); California, USA (solar)

LEDCs:
Sub-Saharan Africa (fuelwood); India (coal and biogas); Tucurui, Brazil (HEP); Rift valley, Kenya (geothermal)

Sample case study: Oil from Alaska (USA)

Location:

The Prudhoe Bay Oilfield is on Alaska's northern (Arctic) coastline.
The pipeline runs south to the port of Valdez.

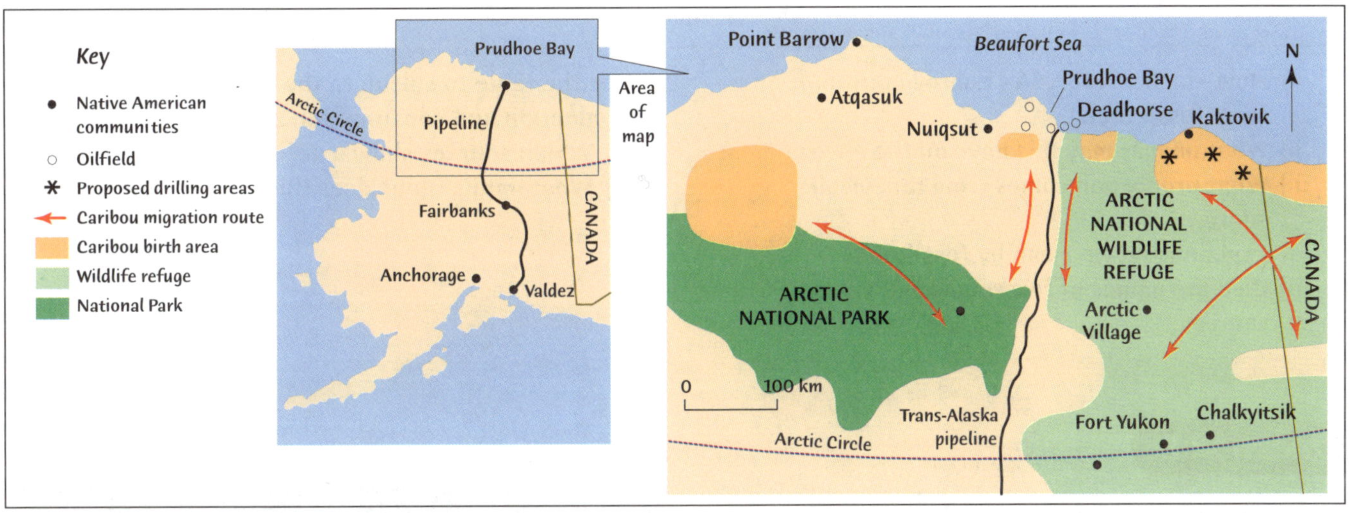

Description:

- oil was discovered in Prudhoe Bay in 1962: the USA's largest oilfield
- construction work began in 1968
- oil had to be piped over 1000km across a mainly frozen wilderness region for export at Valdez
- the first oil flowed through the pipeline in 1977
- in 1989, the *Exxon Valdez* ran aground off Valdez

Benefits of oil production:

- it produces 25% of the USA's oil
- 85% of Alaska's income comes from oil revenue / taxes
- this money pays for schools, hospitals, roads, water supplies and many other services and amenities
- it provides 750 000 jobs across the USA
- oil has given Alaskans the highest average household income in the USA

Problems:

- harsh physical conditions mean production can be difficult
- a major spill would devastate the environment
- the pipeline had to be insulated and built on stilts, three metres above the ground to make it stable
- the route crosses several major rivers, mountain ranges and through an active earthquake zone
- it runs through wilderness regions and fragile ecosystems, crossing the caribou migration route
- some local and Native American people have had to move
- only a small percentage of jobs go to local people

The Exxon Valdez oil spill, 1989

One of the world's worst ever oil tanker accidents occurred when the *Exxon Valdez* spilt 11.2 million gallons of crude oil. It contaminated 1900km of coastline, damaged vast areas of land and marine ecosystems (including three National Parks and four wildlife reserves), devastated the local wildlife and fishing industry, and killed or poisoned ten million birds. The clean-up operation cost $1 billion, plus a further $1 billion in compensation, and is still going on. Oil companies are now seeking permission for new exploration and drilling around Kaktovik (see map).

Exam Practice

Question A

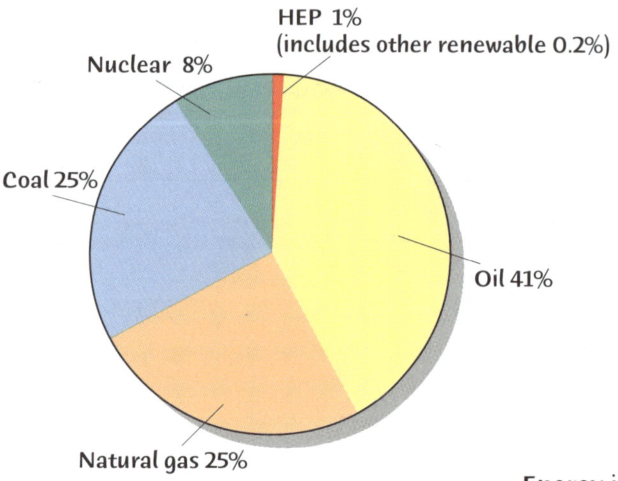

Energy source	Production*	Consumption*
Coal	538	478
Oil	425	774
Gas	456	508
Nuclear	166	166
H.E.P.	26	26

*million tonnes oil equivalent

Energy in the USA

a i) How much of the USA's energy consumption is from fossil fuels?
ii) What impact may this have on the environment?
iii) What proportion comes from renewable sources?

b i) Explain what we mean by fossil fuels.
ii) Give the names of five renewable energy sources.

c Use the table to calculate the USA's energy production and consumption totals.

d i) Which sources show a deficit?
ii) What implications does this have for the USA?

Question B

Look at the photograph.

a What type of energy source does the photograph show?

b Describe in detail:
i) what **conditions** are needed to produce energy;
ii) **how** energy is produced from this source.

c For either wind or tidal power:
i) describe how energy is produced (you may use an annotated diagram);
ii) list the advantages and disadvantages of the energy source.

Question C: case study question

Producing energy can cause major environmental problems. For an energy source you have studied in a named region:

a Name the source and region.

b What benefits has developing the energy source brought to the area?

c What damage has been done to the environment? Explain how this has happened.

d What precautions or measures have been taken to prevent or reduce damage?

Revision Unit 11

Development

The Check up

1 Wealthy, developed countries are called __________ (MEDCs), countries which are less wealthy and developed are called __________ (LEDCs).

2 The majority of the world's LEDCs are found in the __________.

3 A high number of people per doctor is more likely in __________.

4 Employment structure is a good indicator of development – high numbers in __________ jobs is more likely in less developed countries.

5 Data and measurements used to compare development levels are called __________.

6 LEDCs tend to trade in __________ value goods like __________ materials.

7 Look at table (a). Rank the three countries in order of development, with the most developed first. 1 = __________; 2= __________; 3 = __________.

	A	B	C
GNP	350	17 830	1990
Birth rate	31	13	27
Infant mortality	88	7	36
Life expectancy	60	76	70
Adult literacy	57	99	90

Table (a)

8 __________ is the average age somebody can expect to live to.

9 Infant mortality is the number of babies dying before their __________ birthday.

10 __________ population growth rates are more common in MEDCs.

11 Food intake (diet) is measured in __________.

12 __________ (GNP) is the basic measure of a country's wealth, measured in __________.

13 HDI stands for __________, a more complete measure of development than GNP.

14 HDI is calculated from GNP, education data and __________.

15 Unlike GNP, HDI can show differences in development within __________.

16 Some factors used to compare development cannot be __________ easily.

17 The Rostow Model shows __________ stages of economic development.

18 Countries like Japan, the UK and the USA are at stage __________ with high numbers in __________ industry.

19 __________ term aid is needed to deal with sudden emergencies.

20 __________ (NGOs), e.g. charities, are major donors of aid to LEDCs.

Mark yourself *(see answers on page 101)*

- More than 12? Do the Exam Practice – or go on to another unit.
- Less than 12? Do the Crash Course on the next page.

/20

The Crash Course

Development

Development can be measured by looking at wealth. If a country is wealthy (rich) we say it is more developed – an MEDC (More Economically Developed Country). A country which is not wealthy (poor) is less developed or developing – an LEDC (Less Economically Developed Country).

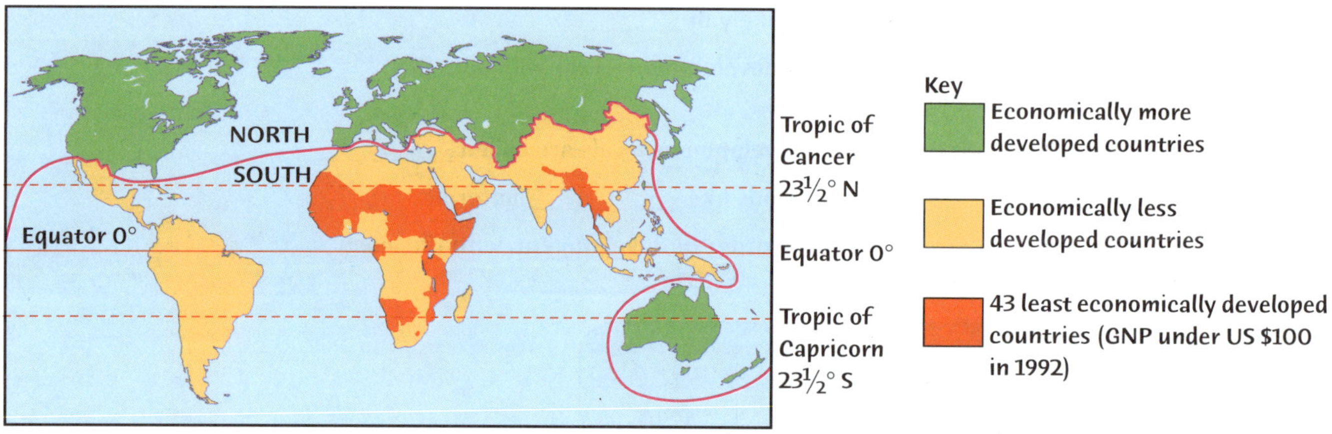

The map is shaded and divided into MEDCs and LEDCs according to GNP. A clear pattern can be seen – countries in the south are mainly LEDCs, whilst those in the north are MEDCs. This is sometimes called the **north-south divide**.

Summary of differences in development

Indicator	Less Economically Developed Country	More Economically Developed Country
Employment (jobs)	High % in primary sector, mainly farming; small % in secondary and tertiary sectors	Highest % in tertiary sector, followed by secondary; low % in primary sector
Population	High birth rate and high but falling death rate = high growth rate; lower life expectancy	Low birth and death rates = low growth rate; high life expectancy
Housing	Often poor quality; lack basic amenities e.g. water, sewage disposal and electricity; millions live in shanty towns on edges of urban areas	Good standard – good range of basic and other amenities
Health	High infant mortality; large numbers of patients per doctor; few nurses, clinics and hospitals, especially in rural areas; more disease (poverty)	Good health provision – low infant mortality, small numbers of patients per doctors; many nurses, clinics and hospitals
Diet	Often poor – average 2000 calories per day; mainly carbohydrate, little protein	High calorie intake, over 3500 per day; wide variety of food available; high protein
Education	Low literacy rates (especially amongst girls and women); reasonable primary school provision, often low % in secondary school	Full-time compulsory primary and secondary education; high literacy rates (no gender difference)
Trade	Low volume and value; mainly primary products (food and raw materials); low value = low profit; imports are expensive (from MEDCs)	High volume and value; mainly manufactured goods (cars, electrical goods); imports often low-cost raw materials (from LEDCs)

All the information in this table can be measured, the data compared and judgements made as to development levels of individual countries. The table covers a range of data, not just economic or wealth.

The Crash Course

Development indicators

Data and measurements used to compare development levels of countries are called **development indicators**.

1994	GNP (US $)	Birth rate (per 1000)	Death rate (per 1000)	Infant mortality rate (per 1000)	Population under the age of 15 (%)	Life expectancy (years)	People in primary jobs (%)	Energy consumption per person	Urban dwellers (%)	People per doctor	Adult literacy (%)
Bangladesh	220	41	14	108	44	53	59	8	19	7810	36
Brazil	3 020	26	8	57	36	66	25	81	79	1660	81
Egypt	660	31	9	57	40	61	41	70	45	970	49
India	290	31	10	88	36	60	62	35	28	3690	50
Japan	31 450	12	8	5	19	79	7	474	78	780	99
Kenya	270	47	10	64	50	61	81	11	28	7870	75
UK	17 970	14	12	8	19	76	2	540	90	650	99
USA	24 750	14	9	8	21	76	2	1074	78	520	99

Explanations of main indicators

Indicator	Description	Measured in / by units	MEDC or LEDC
GNP	Total value of goods and services produced by a country	$ per capita (per person)	high = M low = L
Birth rate	Number of babies born	Per 1000 of total population per year	high = L low = M
Death rate	Number of people dying	Per 1000 of total population per year	high = L low = M
Infant mortality	Number of infants dying before first birthday	Per 1000 of total population per year	high = L low = M
Population under 15	Proportion of the population under 15 years of age	%	high = L low = M
Life expectancy	Average number of years a person can expect to live to in any country	Years	high = M low = L
People in primary jobs	Proportion of working population in primary jobs / work	%	high = L low = M
Energy consumption	Amounts of energy used	Million tonnes oil equivalent	high = M low = L
Urban dwellers	Proportion of total population living in towns and cities	%	changing
People per doctor	Total number of people per doctor	Ratio (e.g. 1:2000) - population divided by total number of doctors	high = L low = M
Adult literacy	Proportion of adults who can read and write	%	high = M low = L

THE CRASH COURSE

Gross National Product (GNP)

- GNP is the **total value** of goods produced and services provided by a country.
- It is usually measured **per capita** – the value is divided by the total population of the country to give a value per person.
- GNP is given in **US dollars ($)** so that accurate comparisons can be made.
- The figure does not mean that every person earned or received that amount of money.
- It is what every person **would** have received if the GNP for their country was **evenly** shared out.
- There are huge differences in GNP across the world – and in wealth in individual countries.
- There seem to be close links between **GNP and development**: highest in MEDCs and lowest in LEDCs.

GNP is an important indicator or measure of wealth and **economic development**. Although wealth is a major factor in the level of development of a country, relying exclusively on economic data can give an inaccurate picture. Other indicators are also important.

Human Development Index (HDI)

Rank order (GNP)	Country	Life expectancy at birth (years)	Adult literacy rate (%)	Years of schooling (average)	GNP per capita ($)	HDI	HDI rank
1	Japan	78.6	99.0	10.7	17 616	0.940	8
6	USA	75.9	99.0	12.3	21 449	0.943	4
10	UK	75.7	99.0	11.5	15 804	0.932	14
70	Brazil	65.6	81.1	3.9	4 718	0.809	62
124	Egypt	60.3	48.4	2.8	1 988	0.612	112
129	Kenya	59.2	49.1	2.7	1 921	0.463	137
134	India	59.1	48.2	2.4	1 072	0.451	139
147	Bangladesh	57.8	35.3	2.0	872	0.371	147

Since 1990 the United Nations has combined a range of data to measure human development. By taking figures for life expectancy and adult literacy combined with average years of schooling and real GNP per capita, the **Human Development Index (HDI)** can be calculated. Each of the separate indicators is given a score of between 0 and 1.0. The final HDI is the average score. Remember that:

- recording data and taking measurements allows objective comparisons to be made;
- figures for GNP and HDI can broadly identify levels of development;
- many other factors contribute to overall quality of life;
- some factors cannot be measured so easily e.g. freedom, human rights, security and happiness;
- HDI can be used to look at areas **within** countries to identify differences and inequalities e.g. between ethnic groups, gender etc.;
- when the HDI score is compared to GNP, a number of countries are ranked differently;
- the **overall** pattern and rankings for GNP and HDI are broadly similar.

THE CRASH COURSE

Stages of economic development

Economic development is a complex process to understand. The **Rostow Model** shown here tries to simplify the five different stages economies can go through.

1 ***Traditional subsistence economy:***
Little money – mainly farming, little or no processing

2 ***Pre-conditions for economic take-off:***
Financial aid and investment, industries and transport develop

3 ***Take off and sustained growth:***
Rapid economic growth, more money and technology. Core region develops

4 ***The drive to maturity:***
Economic growth expands, rapid urbanization, decline in primary industry, rise in secondary industry

5 ***The age of high mass consumption:***
Tertiary industry expands, less secondary industry; population growth slows down

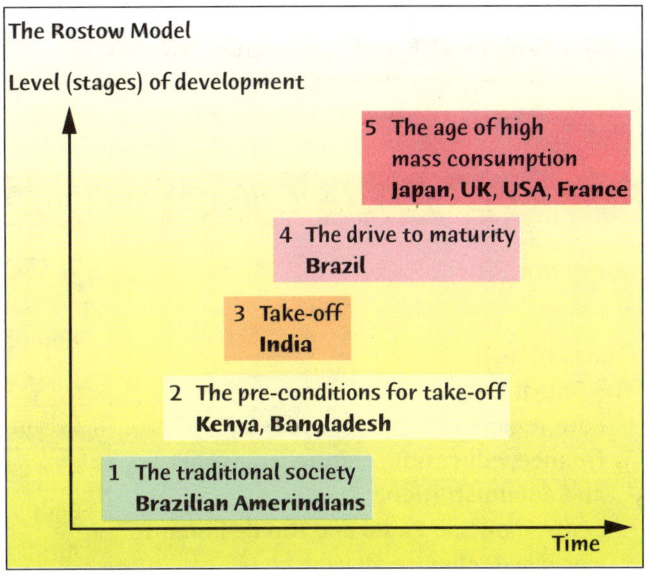

Rostow argued that for LEDCs to develop their economies, they have to pass through each stage of the model and need a high level of extra money or investment to do so. However, money on its own is not enough to ensure economic development: resources, raw materials and relevant technology are all important factors.

Aid

Aid is help or assistance given to a country. Most MEDCs give a range of aid to LEDCs. This may include money, equipment, food and / or advice. Aid can be:

- **short term** – also called emergency aid, usually given to deal with a crisis e.g. food and water shortages, the aftermath of a natural disaster
- **long term** – to help improve the overall quality of life, building infrastructures e.g. digging a well (not supplying bottles of water), training and teaching skills
- **bilateral** – government to government (direct)
- **multilateral** – from governments via agencies e.g. UNICEF
- **non-governmental** – from non-governmental organisations (NGOs) e.g. charities or other groups like Oxfam, ActionAid
- **government** – official development assistance.

Benefits of aid to LEDCs:

- can be large or small scale, country-wide or specifically targeted
- can improve quality of life
- can develop infrastructure and industry
- can create jobs and improve the economy.

Disadvantages of aid to LEDCs:

- may not reach the right people
- not enough for everyone
- may be given as loans, leading to debt
- may be unsuitable, unsustainable or inappropriate e.g. large prestigious projects when smaller projects may be what is needed.

The Case Studies

Examples

Commonly used case studies:
MEDCs:
UK; France; Italy; USA; Japan
LEDCs:
India; Bangladesh; Kenya; South Africa; Brazil; Mexico

Most development case studies look at and compare and contrast countries or regions at different stages of development, looking at development indicators, quality of life and economic activity.

Sample case study: Regional development in Brazil

Location:

See map

Description:

The South East

- core region – centre for industry, finance, education, culture, transport and administration
- centred on São Paulo and Rio de Janeiro
- warm, wet climate all year
- important farming region e.g. coffee, sugar cane
- mineral resources e.g. iron ore, gold and diamonds
- major ports developed for import and export of goods
- good energy resources – HEP (local) and oil (imported and refined in Rio)
- well-developed secondary industry e.g. cars, ships, food processing
- a range of jobs (many well paid) available
- widespread government and foreign investment
- highest standard of living in Brazil

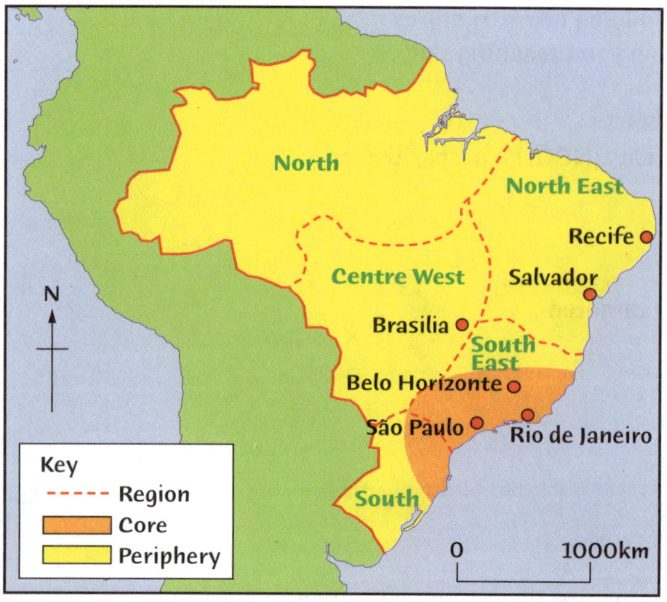

Brazil

Brazil	South East	North East
Population density	63 per km²	25 per km²
Birth rate (per thousand)	22	48
Infant mortality rate (per thousand)	49	109
Life expectancy (years)	63	48
People per doctor	875	2150
Adult literacy (%)	72	39
Brazil's industrial production (%)	56	16
People employed in industry (%)	70	10
Brazil's energy consumption	71	13
Urban dwellers (%)	84	47
With electricity (%)	82	15
Clean water (%)	64	23
Car ownership (%)	66	10

Datafile on Brazil

The North East

- periphery (edge) region, remote from core in south east
- main coastal cities are Salvador and Recife
- hot and dry climate, unreliable rainfall can lead to drought
- majority of people are subsistence farmers or landless, working on other people's land
- soils are poor following years of drought and over-use
- some plantation crops e.g. sugar cane, cocoa grown for export
- few natural resources – one HEP station insufficient for needs
- lack of investment = little economic development – poor industry, transport, and services
- thousands have moved away, many migrating to cities in the south

This case study reinforces the point that there is not only a great difference between levels of economic development from country to country, but also great contrasts within countries. There are few countries (even MEDCs) that don't have regional differences in wealth and quality of life.

EXAM PRACTICE

Question A

Look at Table (a).

a i) List the countries you consider to be MEDCs.
  ii) Give reasons for your choices.

b Countries are often classified as either MEDCs or LEDCs. Looking at the table, why could this be misleading?

c Describe and explain the main factors affecting life expectancy.

d i) Name three other indicators which could be used to compare development.
  ii) Write a sentence for each one explaining what it measures.

Table (a)

Country	Variables		
	GNP (US $)	Life expectancy (years)	Birth rate (per thousand population)
Bangladesh	220	53	41
Brazil	3020	66	26
Egypt	660	61	31
India	290	60	31
Japan	31 450	79	12
Kenya	270	61	47
UK	17 790	76	14
USA	24 750	76	14

Question B

Part 1:

a i) What is GNP a measure of?
  ii) How is it calculated?

b Using the data in Table (a), draw two scattergraphs to show:
  i) GNP and life expectancy
  ii) GNP and birth rate.
  (GNP should be plotted along the horizontal axis).

c For each graph, describe whether there is a positive or negative relationship or correlation.

Part 2:

a i) How is HDI measured?
  ii) Why is it considered a more useful indicator of development than GNP?

b Study the map.
  i) Name three countries with HDIs over 0.90.
  ii) Name three countries with HDIs under 0.499.

c Describe the general pattern of HDI distribution shown by the map.

World HDI rates

Key

- 0.90 and over (32 countries)
- 0.75–0.899 (32 countries)
- 0.50–0.749 (33 countries)
- 0.25–0.499 (30 countries)
- 0.048–0.249 (31 countries)
- No data

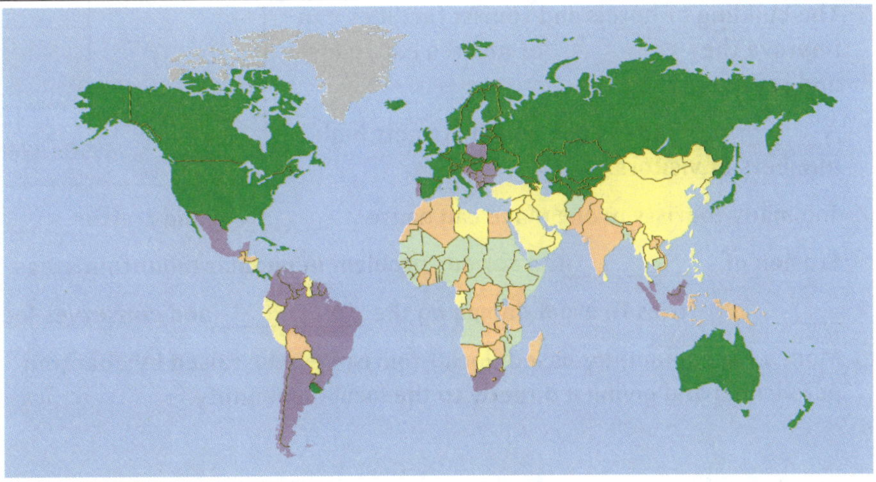

Question C: case study question

For a named country you have studied, compare and contrast two different regions.

a Draw a sketch map of the country showing the location of the two regions.

b Use the following headings to describe each region: climate; industry; health; population.

c Explain why there are differences between the two.

Revision Unit 12

Leisure and tourism

The Check up

1 People have far more free time today, working __________ hours and having longer holidays.

2 Free time is usually called __________ time.

3 Tourism is classed as a __________ industry, providing a __________.

4 __________ holidays are all-inclusive deals including the cost of flights, accommodation and meals.

5 Travel within your own country is called __________ tourism.

6 Travel from one __________ to another is called international tourism.

7 Backpackers travelling the world are often called __________ tourists.

8 __________ holidays involve travelling to distant places e.g. Australia.

9 Tourism is the world's __________ growing industry.

10 Study graph (a), then complete the sentence.

Between 1970 and 1990 the number of tourists increased by __________.

Graph (a) growth in world tourism

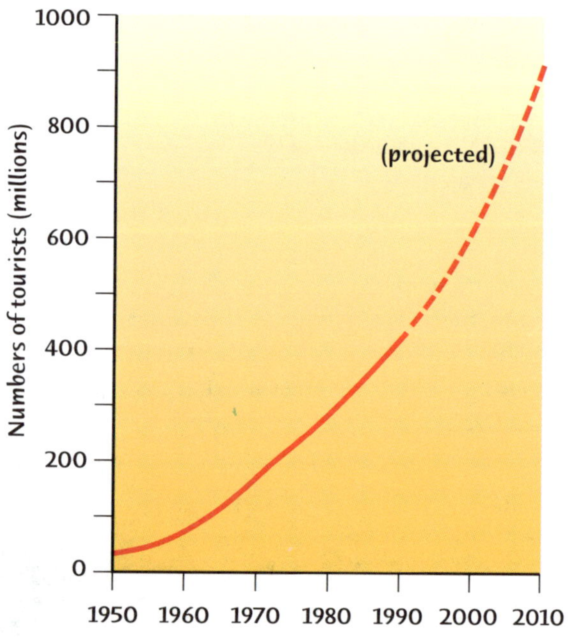

11 Tourism is an important part of the economy in many countries, but it can cause a major impact on the __________.

12 Although tourism may bring many jobs, they are often unskilled, __________ paid and __________.

13 Tourists bring valuable __________ currency to LEDCs.

14 __________ can occur between the needs of tourists and local people.

15 The building of hotels and tourist facilities can improve the __________ of an area e.g. roads and services.

16 A __________ site is one which attracts high numbers of visitors.

17 Too many tourists in one place can cause __________ and traffic __________.

18 Erosion of __________ is a serious problem in popular mountain areas.

19 __________ tries to avoid damaging the __________ and conserves local resources – it is sustainable.

20 Money for community-based (local) tourism can be raised by charging __________ (a certain amount per visitor) and giving it directly to the local community.

Mark yourself *(see answers on page 101)*

- More than 12? Do the Exam Practice – or go on to another unit.
- Less than 12? Do the Crash Course on the next page.

/20

THE CRASH COURSE

Leisure and tourism

Most people in MEDCs today have far more leisure time than 30 years ago. People choose how to spend their free time, some on passive activities like watching TV and others in more active pursuits e.g. sports. Increasingly, people spend large amounts of leisure time on holiday or visiting places away from home – as visitors or tourists.

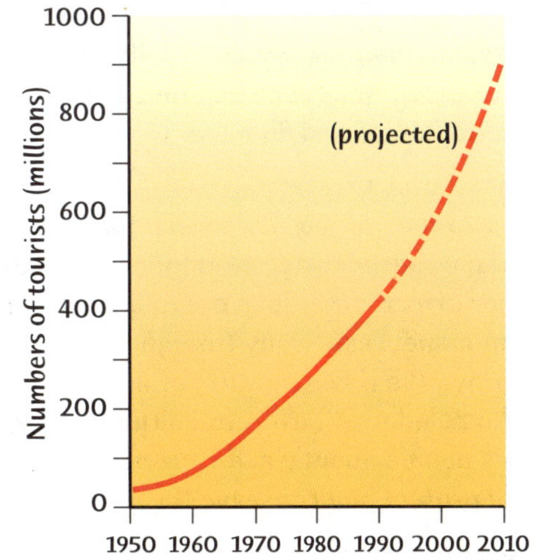

Tourism

Tourism is a service industry, so it is classed as a tertiary industry. It is the world's fastest-growing industry because:

- many people work shorter hours and have paid holiday entitlement
- increased mobility – a high proportion of people own cars
- air travel has made the world 'shrink' – people can travel further more quickly
- package holidays offer flights and accommodation at affordable prices
- people are living longer, and many are retiring early – there is a growing number of active older people
- it currently employs about 10% of the world's workforce.

KEY FACTS TO LEARN ABOUT TYPES OF TOURISM

- **Domestic**: within your own country
- **International**: to another country
- **Mass**: aimed at large numbers of tourists, usually via package deals
- **Independent**: typically younger people, often students taking a gap year, travelling around the world e.g. backpackers, perhaps working on the way to fund their travel
- **Day trips**: short visits to local attractions e.g. theme parks, activity centres, areas of attractive landscape or scenery, seaside resorts or heritage, cultural and historical sites
- **Weekend / short breaks**: similar locations to day trips, but time to travel further – often at special low-price deals
- **Package holidays**: cheap all-inclusive one or more week holidays; includes travel, accommodation and meals, typically to warm, sunny destinations
- **Specialist**: e.g. activity holidays, safari, winter sports, cruise, climbing, historic or cultural
- **Long-haul**: long-distance travel e.g. from the UK to Australasia, Thailand and the Far East, Antarctica, Pacific Islands

THE CRASH COURSE

The impact of tourism

Tourism is expanding rapidly, especially in LEDCs, helping the economies of many countries to grow. However, sometimes economic gains can be offset by environmental losses and damage.

Advantages of tourism:

- provides a range of jobs for local people
- jobs bring wages: money spent locally and taxes paid
- multiplier effect: other jobs created e.g. food supply to restaurants
- brings in money: especially foreign currency to LEDCs
- can improve the infrastructure of an area
- brings knowledge of different cultures and ways of life to wider audience
- educates people about e.g. different environments and conservation
- can help protect and conserve the environment, historical sites etc.

Disadvantages of tourism:

- jobs are often unskilled and poorly paid e.g. cleaners, kitchen staff
- managers are often brought in from abroad
- work may be part time and / or seasonal
- profits may leave the country or area (e.g. hotels owned by foreigners)
- local people may lose land to new developments
- fragile ecosystems may be damaged by large numbers of visitors e.g. coral reefs
- tourists may cause overcrowding, traffic congestion, litter and pollution
- other economic activities may suffer from disturbance e.g. farming
- scarce resources e.g. water may be used up by tourists, not available for local people.

It is easy to see how conflicts can occur as the needs and demands of tourists, tour companies, local people and the environment can all be quite different. Tourism works best when local people are consulted over developments affecting their community.

Honeypot sites

These are sites (often small) e.g. historic buildings, mountain peaks, which attract large numbers of tourists. The sites themselves can become damaged or spoilt e.g. erosion of footpaths, general congestion. It is difficult to maintain such sites, while allowing large numbers of people to visit them.

Why honeypot sites attract an increasing number of visitors

THE CRASH COURSE

Sustainable tourism

Tourism is the world's fastest-growing industry, yet many countries are concerned about the negative effects of mass tourism. Tourism contributes significantly to the economies of both MEDCs and LEDCs. In some LEDCs, developing the tourist industry may be the only way to significantly improve their economies. To avoid problems the industry must be managed properly – **sustainable tourism** is considered by many to be one way of balancing different needs and reducing conflict. It seeks to:

- avoid damaging the environment
- integrate developments to fit in with existing plans
- involve local communities in planning
- consider how local cultures and customs contribute to tourism
- support small-scale projects, decreasing dependence on TNCs
- conserve resources used by the tourist industry by recycling and minimising waste.

Footpath erosion at Mill Hill on the Pennine Way, Derbyshire

There has been a great effort by tour companies, governments and local communities to make sustainable tourism succeed. A World Conference held in 1995 set out to establish clear codes of conduct to deal with pollution and to encourage tourists to learn more about the cultures and environments of the places they visit.

Ecotourism

Ecotourism was an early form of sustainable tourism started by a number of tour companies taking small groups on specialized holidays e.g. on wildlife safaris or to remote, fragile environments like Antarctica or the Amazon rainforest. Originally it attracted wealthy tourists in search of something different, but today it has a wider appeal – but is not aimed at the mass market.

Ecotourism today:

- is on a small scale, organized closely with the consent of the local communities affected
- directly benefits local people and communities e.g. via levies on each visiting tourist
- directly benefits local wildlife e.g. villagers will look after animals in their area (not kill them) as they know they can then attract tourists to see them
- is developed as a partnership between tourists, tour operators and local people
- can also be called community-based tourism.

Protection, not poaching … wildlife attracts tourists

CASE STUDIES

EXAMPLES

Commonly used case studies:
UK:
Local studies e.g. seaside resorts (Blackpool); National Parks e.g. Lake District, Snowdonia;
MEDCs:
Florida (DisneyWorld); Everglades and Yellowstone (National Parks) (USA); Mediterranean resorts; winter sports (Alps); Disneyland, Paris (France)
LEDCs:
Caribbean; Kenya; Zimbabwe; South Africa

Sample case study: Ecotourism in Namibia

Location:

Namibia is on Africa's south-west coast, north of South Africa and south of Angola.

Namibia

Description:

Like many African countries, Namibia relies on tourism to bring in much needed money – and about 10 000 jobs. Each year, 300 000 visitors come to visit the dozen or more National or Game Parks and the spectacular desert and coastal scenery. With the number of jobs and visitors estimated to rise to 50 000 and 600 000 respectively, the government considers this sector essential for Namibia's economic development. It has also had to consider the impact so many visitors might have on what are often fragile ecosystems.

Ecotourism is a high priority with Namibia which is keen to:

- encourage high quality ecotourism
- restrict numbers in fragile areas
- keep a balance between the needs of tourists and conservation of the environment
- develop community-based projects so tourists can see the 'real' Namibia
- educate local people about the importance of tourism as a means of preserving the wildlife and the environment and of making money.

A good example of this approach can be found at Purros in the north of Namibia. Local people used to kill elephants in the area. Now they actively protect the elephants in return for levies from two tour operators who bring small groups of tourists to see the animals in their natural environment. The local people also raise money from selling souvenirs. Tourists, tour companies, the local people and the wildlife benefit from the project, but a delicate balance has to be kept between their various needs for the project to remain sustainable. A major threat is from encouraging too many visitors and from visitors who are not part of official groups.

EXAM PRACTICE

Question A

Part 1:
Look at the photograph which was taken in Bowness (on Lake Windermere) in the Lake District National Park.

a This is typical of a small area which attracts a lot of people. What do we call such sites?

b Describe and explain some of the problems associated with these sites.

Traffic congestion in Bowness, Cumbria

Part 2:
Study the pie chart showing land ownership in the UK's National Parks.

a Why is the term 'National Park' misleading?

b Each National Park receives millions of visitors every year. Describe and explain three different conflicts of interest which might occur between the groups represented on the graph and tourists.

Who owns the National Parks?

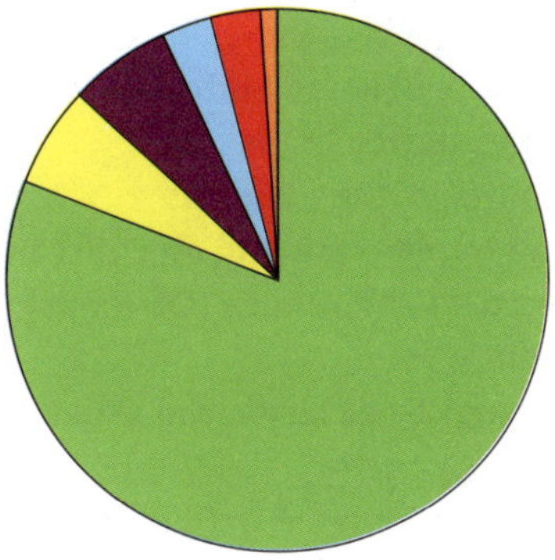

Question B

Study the table showing popular new holiday destinations for UK tourists over the past 20 years.

Caribbean	Africa	South East Asia	Australasia
Jamaica	Egypt	India	Australia
Barbados	Kenya	Sri Lanka	
St Lucia	South Africa	Thailand	
	Zimbabwe	Singapore	

Holiday destinations for UK tourists, 1975-95

a i) Name two factors which most of these locations have in common.

ii) What name is given to this type of tourism?

b Explain why these locations have become important for UK tourists in the past 20 years.

c What benefits might tourism bring to these countries?

Question C: case study question

For a named tourist destination you have studied:

a Name and locate the destination (you may draw a sketch map).

b Describe and explain why tourists visit the area.

c How has tourism affected:

i) the environment

ii) the economy

iii) the local people?

Revision Unit 13
Map skills

General information

Maps are an essential part of geography. Throughout your GCSE course you will be dealing with a range of different maps. Your final exam papers will contain several maps - these may be:

- to give you information e.g. site and situation / location of a place
- part of a question e.g. complete the map
- sketch maps, labelled or annotated maps, Ordnance Survey (OS) maps
- maps at different scales covering small or large areas
- specialized maps e.g. choropleths, weather maps.

You will be expected to show that you have a range of map skills. You should be able to:

- accurately find and extract requested information from various maps
- draw and annotate simple sketch maps
- give directions using points of the compass
- use the scale of the map to measure and give distances and area
- give four- and six-figure grid references (OS maps)
- describe a route in detail
- use contour lines (OS maps) to describe the relief of the land and draw cross-sections
- identify physical features and human activity e.g. river features or types of settlement.

Ordnance Survey (OS) maps

You will almost certainly find that an OS map extract (like the one on p.92) is included on at least one of your final exam papers. It may be part of a compulsory question (no choice - you have to do it) or one of a range of questions you have to choose from. Check with your teacher which is the case BUT, for most GCSE candidates, assume that you will have to answer a question based on an OS map.

In many ways this is good - you will have the map in front of you, and usually the key to the symbols.

Although you don't have to know or learn all the symbols, it does save a great deal of time (especially in the exam when time is an important factor) if you are already familiar with the main ones. This shouldn't be a problem since you will have used OS maps throughout secondary school (not just during your GCSE course).

So what do you need to know? The following is a summary of the range of map skills you may be tested on in your exams, together with some practice exercises you can do using the map extract (p.92).

Direction

Whenever you describe the location of something or a direction in which you would have to travel on a map you should use compass directions. For example, 'the M5 motorway is located to the east of Bridgwater, running north to south'; 'Cannington village is north west of Bridgwater'.

Exercise 1

In which direction would you be travelling if you were going from:

a Bridgwater to North Petherton;
b Pawlett to Puriton;
c Durleigh Reservoir to Cannington?

Tip: 'Travelling from' - start at the <u>first</u> location and state the direction you would have to travel to get to the <u>second</u> location (don't get it the wrong way round).

Information and exercises

Scale, distance and area

Most OS maps used in exams are either 1:50 000 (2 cm to 1 km) or 1:25 000 (4 cm to 1 km). In both cases, each grid square is 1 km^2. Both scales have different sets of symbols and cover all of Great Britain. 1:50 000 maps cover a larger area, but in less detail, than 1:25 000 maps.

Tip: When you are working out area, count up the number of **full** squares (each one is 1 km^2), then add on the **parts of** squares to get the total.

Exercise 2

a How long is the section of the M5 motorway shown on the map?

b How far is it in km from Rodway Farm (256407) to Copse Farm (314335)

 i) in a straight line;

 ii) by road?

c What area does Bridgwater occupy (square kilometres)?

d How big is The Thickets wood?

Grid references

You will need to be able to look up and give both four-figure grid references and six-figure grid references. Make sure that you read the question carefully – you may lose marks if you give a six-figure grid reference when asked for a four-figure one and vice versa. A four-figure grid reference is used to give a square or general location e.g. a woodland area or lake; a six-figure grid reference is used to give a specific location e.g. an individual building. Always give the number of the northing (the vertical line which runs north to south) first, then the easting (the horizontal line running east to west).

Exercise 3

a What is the main feature in each of the following squares?

 i) 3032 ii) 2539 iii) 2636

b What feature is at:

 i) 266410 ii) 304341 iii) 318444?

c Give a four-figure grid reference for:

 i) the factory north east of Puriton

 ii) Combwich village

 iii) Cannington Brook.

d Give a six-figure grid reference for:

 i) the telephone box in Stretcholt

 ii) the post office in Goathurst

 iii) Pawlett Hill.

Describing routes

Describing a route on a map is an important skill. You need to include enough information to describe it well – but not so much that it is too detailed.

Exercise 4

Complete the following description of the route from Lovedere Farm (257353) to the agricultural college in Cannington (254397).

a Travelling ________ from Lovedere Farm under the ________ line, turn ________ at the first T-junction into the village of ________. Turn left out of the village, past the ________ on the left. Where the road forks, turn ________ until you reach the main road (the A ________). Turn left onto this road, travel in a ________ direction for about 2 km, passing ________ farm on your right. At the roundabout, go ________ across into Cannington. At the T-junction turn ________, and then first ________. The college is on your ________.

b Using (a) as a guide, describe the route from the landfill site (313430) to West End Farm (333380).

OS map extract

Reproduced from the OS map with the permission of the Controller of Her Majesty's Stationery Office, © Crown copyright

Communications

ROADS AND PATHS

Not necessarily rights of way

- Service area, M 1, Elevated, Junction number 1 — Motorway (dual carriageway)
- Motorway under construction
- Unfenced, Footbridge, A 470 (T) — Trunk road
- Dual carriageway, A 493 — Main road
- Main road under construction
- B 4518 — Secondary road
- A 855, B 885 — Narrow road with passing places
- Bridge — Road generally more than 4m wi
- Road generally less than 4m wid
- Other road, drive or track
- Path
- Gradient : 1 in 5 and steeper; 1 in 7 to 1 in 5
- Gates — Road tunnel
- Ferry P — Ferry (passenger); Ferry V — Ferry (vehicl

PUBLIC RIGHTS OF WAY

- Footpath
- Road used as a public path
- Bridleway
- Byway open to all traffic

OTHER PUBLIC ACCESS

- Other route with public access
- National Trail, Long Distance Route, selected Recreational Paths
- Selected cycle route/trail
- 8 National Cycle Network number

Danger Area — Firing and Test Ranges in the area.
Danger! Observe warning notices.

RAILWAYS

- Track multiple or single
- Track narrow gauge
- Bridges, Footbridge
- Tunnel
- Viaduct
- a — Station, (a) principal
- Freight line, siding or tramway
- LC — Level crossing

General Information

LAND FEATURES

- Electricity transmission line (pylons shown at standard spacing)
- Pipe line (arrow indicates direction of flow)
- ruin — Buildings
- Public building (selected)
- Bus or coach station
- Place of Worship: with tower; with spire, minaret or dome; without such additions
- Chimney or tower
- Glasshouse
- Heliport
- Triangulation pillar
- Radio or TV mast
- Windpump/wind generator
- Windmill with or without sails
- Graticule intersection at 5' intervals
- Quarry
- Spoil heap, refuse tip or dump
- Coniferous wood
- Non-coniferous wood
- Mixed wood
- Orchard
- Park or ornamental ground
- Forestry Commission access land
- National Trust-always open
- National Trust-limited access observe local signs
- National Trust for Scotland

BOUNDARIES

Administrative boundaries as at **September 1997**

- National
- District
- County, Unitary Authority, Metropolitan District or London Borough
- National Park or Forest Park

WATER FEATURES

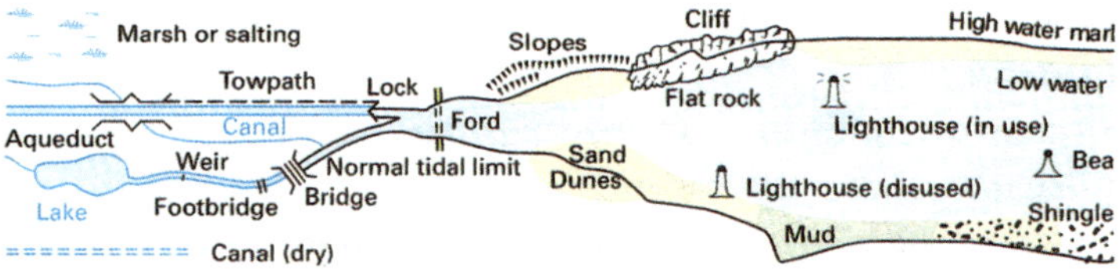

Contour values in lakes are in metres

ARCHAEOLOGICAL AND HISTORICAL INFORMATION

+	Site of monument	⚔	Battlefield (with date)	VILLA	Roman
• ∘	Stone monument	☆ ····	Visible earthwork	Castle	Non-Roman

Information provided by the Royal Commissions on Historical Monuments for England and Ancient and Historical Monuments for Scotland and Wales

ABBREVIATIONS

CH	Clubhouse	CG	Coastguard
MS	Milestone	P	Post office
PC	Public convenience (in rural areas)	MP	Milepost
TH	Town Hall, Guildhall or equivalent	PH	Public house

HEIGHTS

- 50 — Contours are at 10 metres vertical interval
- ·144 — Heights are to the nearest metre above mean sea level

Heights shown close to a triangulation pillar refer to the ground at the base of the pillar and not necessarily to the summit

ROCK FEATURES

Tourist Information

TOURIST INFORMATION

- Information centre, all year / seasonal
- Viewpoint
- Parking
- Picnic site
- Camp site
- Caravan site
- Selected places of tourist interest
- Telephone, public/motoring organisation
- Golf course or links
- PC — Public convenience (in rural areas)
- Youth hostel

Information and exercises

Relief

Relief is the height and shape of the land. Height is shown in three ways:

- spot heights – which give the precise height shown by a black dot and number e.g. at 287406;
- triangulation pillars – posts, usually on hill tops, shown by a blue triangle with a central blue dot plus black number e.g. at 292431;
- contour lines – brown lines joining up points of equal height. The closer together they are, the steeper the slope e.g. in square 3240. If there are very few or no contour lines the land is flat e.g. in square 2640.

On OS maps height is shown in metres. Contour lines are drawn at 10 metre intervals. The shape of the landscape is shown by the shape of the contour lines. You need to identify the basic, common shapes e.g. hill, ridge, valley, plateau, escarpment. Most flat areas are lowland (in the absence of contour lines look for spot heights), although plateau-like highland areas are flat at the top.

You may be asked to draw, complete or label a cross-section from an OS map. A cross-section shows the height and shape of the land in detail, and can help to identify distinctive physical features e.g. a river valley.

Exercise 5

a What is the height at: i) 272349 ii) 263387 iii) 272331?

b Which land is steepest: square 2943 or square 2733?

c How would you describe the relief of the land in squares 2841 and 2941?

d Which area of the map has the highest land: the north west, the south east or the south west?

Physical features

Exam questions often ask you to identify physical features such as:

- coastal landforms e.g. cliffs, headlands, bays, beaches, spits (Unit 4, p.29);
- rivers and river valleys e.g. waterfalls, gorges, meanders, embankments, estuaries (Unit 2, pp.16-17);
- mountain features e.g. glacial landforms (arêtes, corries, U-shaped valleys) (Unit 3, p.23);
- landforms associated with named rock types e.g. chalk escarpment, Carboniferous limestone features (clint pavements, disappearing streams, swallow holes).

Learn how to recognize these features on an OS map – a question may ask you to give map evidence (a grid reference) to identify such features. The map extract on page 92 is of a lowland area – the main physical feature is the River Parrett flowing from the south east to the north west.

Exercise 6

a What evidence is there that the River Parrett flows north west?

b What river features are found at: i) 275410 ii) 307404 iii) 273403 iv) square 2840?

c Describe the differences between the River Parrett in squares 2643 and 3234.

Human activity

You will be expected to be able to identify a range of human activities on a map. These may include:

- settlement e.g. distribution, site, types and functions (Unit 8, pp. 53-4);
- transport e.g. distribution, types (road, rail, water);
- economic activity e.g. farming, industry (Unit 9, pp. 63-5), tourism (Unit 12, pp. 85-7).

Exercise 7

a Which type of settlement is:
   i) Bridgwater
   ii) Horsey (327392)
   iii) Pawlett (300428)?

b i) Describe the route taken by the M5.
   ii) How and why is this slightly different to the route taken by the railway?

c Give map evidence for four different types of economic activity.

Information and exercises

Answers to map skills exercises

Exercise 1
a south
b south east
c north west

Exercise 2
a 14 km
b i) 9 km ii) 12 km
c 11/12 km^2
d 0.5 km^2

Exercise 3
a i) orchard
ii) Cannington village
iii) Durleigh reservoir
b i) electricity transmission line (pylons)
ii) motorway junction (M5)
iii) level crossing
c i) 3342 ii) 2542 iii) e.g. 2740
d i) 296441 ii) 256342 iii) 292431

Exercise 4
a north east; electricity transmission line; right; Durleigh; reservoir; right; 39; north west; Sandford; straight; left; right; left

Exercise 5
a i) 20 metres ii) 12 metres iii) 111 metres
b 2733
c flat lowland; (flood plain) around 6 metres
d south west

Exercise 6
a The river channel is much wider with many meanders and embankments to the north west - it is much narrower and straighter to the south.
b i) meander
ii) embankment (levée)
iii) tributary (Cannington Brook)
iv) flood plain
c In 2643 River Parrett = wide, slow-flowing, deposition of sand / mud, embankments; in 3234 River Parrett = narrow, fairly straight, valley side rises 10 metres 1km to the west.

Exercise 7
a i) town ii) hamlet iii) village
b The M5 runs north to south in an almost straight line except where it bends to the east around the town of Bridgwater. It follows fairly flat lowland crossing the River Parrett and several tributaries.
c e.g. farms, factory, works, landfill site, golf course, sewage works

Answers

Revision Unit 1: Rocks and tectonics

The Check up answers, p. 7

1 crust, mantle and core; **2** mantle, molten / liquid; **3** plates; **4** igneous; **5** A = destructive; B = subduction; C = volcano; **6** faults; **7** plate; **8** seismograph; **9** Richter; **10** tsunamis, flooding / damage; **11** gas pipes fracture / gas ignites; **12** composite; **13** pyroclastic; **14** has erupted in the last 1000 years; **15** magma, lava; **16** A = shield, B = gentle, C = lava; **17** subduction; **18** Ring of Fire; **19** lava, fertile; **20** weather

Exam Practice answers, p. 12

Question A

a X = destructive, Y= constructive, Z= collision.

b This is a destructive plate boundary.
THINK: 'annotated' means add labels which explain the diagram; e.g. the direction each plate is moving, features caused by the movement. Keep drawings simple e.g. line drawings – use colour only if it helps, otherwise you are wasting valuable time in the exam. Your labels (annotations) need to be concise, with arrows pointing clearly to what is being explained.

c X = This is a destructive plate boundary. Two plates meet here. As one plate slides underneath the other, it melts and forces magma to surface. Y is a constructive margin, gently releasing magma and Z is a collision zone pushing rocks / plates upwards.

Question B

a A = subsidiary cone, B = pipe, C = ash and stream, D = crater, E = layers of ash and lava.

b See the diagram on p. 10.
THINK: 'labelled' means just indicate the main features – NOT explain them.

c THINK: make it clear what the differences are, don't just write lists. For example, a composite volcano has steep sides, is made from layers of ash and lava and is liable to erupt suddenly and explosively, whereas a shield volcano has gentle sides, is made from lava and usually erupts slowly for a longer time.

d Examples: soils near volcanoes are often rich in minerals and are very fertile – good for farming. They may provide the only fertile land in the area. The risk seems worthwhile as a living can be made – and few volcanoes erupt very often.

Question C

This is a case study question – the sample case study on p. 11 (Kobe) could be used to answer this, but you should use the earthquake you have studied.

a There are two parts to this question – location should be both general and specific e.g. in southern Japan and centred on Awaji Island in Osaka Bay near the cities of Kobe and Osaka.
THINK: Sometimes drawing a simple labelled sketch map is more helpful. The second part (easy to miss out) asks why it occurred here. Try to be specific e.g. don't just say 'where tectonic plates meet' – try to remember the names of the plates and in which direction they are moving. Again, a map / diagram might help.

b THINK: primary effects are those which happen immediately e.g. ground shakes so buildings collapse and water mains burst; secondary effects are those which happen later e.g. water mains burst, no water to put out fires (which become out of control) plus water supply becomes contaminated which leads to disease. Long-term damage to mains system, difficult for everyday living and for industry. Costly.

c i) e.g. strict building regulations to make buildings 'earthquake proof'; disaster emergency plans in place to deal with earthquakes.
ii) THINK: what damage was actually done in the example you have used? Were buildings strong enough? Were the emergency services able to cope? If not, why not? What could be learnt to help in the future? (Room for some of your own ideas here.)

Revision Unit 2: Rivers and the water cycle

The Check up answers, p. 13

1 evaporation, precipitation, transpiration, surface run-off (any 2); **2** inputs, stores (or flows); **3** catchment; **4** infiltration; **5** interception; **6** groundwater; **7** load; **8** attrition, hydraulic action; **9** dissolves; **10** deposition, erosion; **11** source, mouth; **12** upper, middle, lower; **13** A = overhanging, B = plunge, C = hard; **14** middle, meanders, ox-bow lakes; **15** A = ox-bow lake, B = embankments or levées, C = silt / sediment / alluvium; **16** shallower, flooding; **17** discharge; **18** heavy, bare, surface, quickly; **19** slowly, interception, permeable, a dam; **20** flood plain

Exam Practice answers, p. 20

Question A

a A = condensation, B = precipitation, C = surface run-off, D = transpiration, E = evaporation.

b **1** – trees intercept rainfall, delaying it from reaching the surface and then rivers – flooding less likely; **2** – tree roots bind the soil together and take up large quantities of moisture from the soil (preventing soil erosion and saturation).

c i) River A: The hydrograph rises steeply, so water reaches the river quickly making it more likely to flood.
ii) Table on p. 18
THINK: make sure when you present your answers it is clear when you are referring to A or B. You must think of your answer as being in two parts – a description and an explanation. For example, hydrograph B (or A) is more likely to flood because the following factor means that water will reach the river quickly, causing it to rise quickly: impermeable rock (water will not sink into it). You need to describe and explain four different factors.

Question B

a i) From south east to north west (ii) a meander (bend).

b THINK: first find the two points given by the grid references. The course of the river is the route it takes – include distance, direction and any notable features in your description e.g. the River Parrett flows north for a kilometre. It is in its middle course – the channel is getting wider. There is a flat flood plain on either side and embankments at the sides of the river. The river then bends sharply (meander) to the east. Continue the description in this way. Make sure you describe both the course and any river features you see.

c You will probably choose a meander or embankment. Make sure you draw your diagrams clearly and that the labels (annotations) explain the formation of the feature. You may need more than one diagram. Don't forget to write down the name of the feature you are describing. Use colour if it helps your answer. (Look at p. 16 if you're stuck!)

d The main land use is farming. There are a number of farms shown – map evidence would mean giving the name and grid reference e,g. Hallicks Farm (Fm = farm) at 289410. Name about three or four. Other evidence might include drainage ditches in square 2741 and the Drain from 340438 to 305450 – both indicate that the land is being drained for use; flood plains are traditionally used for dairy farming. One farm – at 294403 – is actually called Dairy House Farm. Remember – you are only asked to look at the land use in this section (not over the whole map).

Question C

This is a case study question – the sample case study on p. 19 (Mississippi) could be used to answer this, but you should use a scheme / river you have studied. The question asks about river / water management – this may be a named scheme (e.g. Aswan Dam, Nile), part of a river (e.g. River Colorado Cascade) or a whole river basin (e.g. Mississippi).

a The location given should be both general and specific e.g. the Mississippi river basin is in the USA, draining the central third of the country from Canada down to the Gulf of Mexico. As with most locations, a simple labelled sketch map could be drawn instead.

b THINK : the question says describe – NOT explain (yet). Give factual information about HOW (not why) the river is being controlled.

c THINK: this is the 'why' – why did the river / water need to be managed? What were the problems before? For example, the Mississippi – flooded in most years (destroyed farmland and buildings) and is an important transport route (so river channel needs constant dredging to allow large boats to navigate its course).

d THINK: try to describe at least two (preferably three) good effects and three bad effects – the question asks for both.

Revision Unit 3: Ice

The Check up answers, p. 21

1 ice caps; **2** larger / bigger; **3** South; **4** glaciation; **5** hollows; **6** A = freeze-thaw, B = rotational slip, C = abrasion; **7** sticks, moves; **8** tarn; **9** three, fourth; **10** arête; **11** pyramidal peak; **12** V-shaped, deepened, straightened; **13** ribbon lakes, thin; **14** hanging, ice / glacier; **15** dams, reservoirs; **16** deposit / drop; **17** A = recessional moraine, B = lateral moraine, C = medial moraine, D = terminal moraine; **18** recessional; **19** meltwater, outwash; **20** drumlin

Exam Practice answers, p. 26

Question A

a Refer to pp. 22-23.

b **1** – water fills cracks in the rock; **2** – the water freezes into ice and expands, making the crack wider / bigger; **3** – the ice melts, releases the pressure and the rock eventually breaks into pieces. This process is continuous, freezing by night and thawing by day, leading to the gradual break-up of rocks.

c A = corrie, B = arête, C = pyramidal peak. Refer to p. 23 for help. THINK: don't forget to describe the processes at work in your answer.

Question B

a i) Refer to p. 24 – include terminal, lateral, medial and recessional moraine. Moraine is material eroded by a glacier, then deposited (dropped).
ii) Refer to diagram on p.24.

b i)Refer to p. 24. A drumlin is a distinctively shaped hill formed by glacial deposition. You may wish to answer by drawing an annotated diagram.
ii) The ice has moved from the steeper side towards the gently sloping side.

Question C

This is a case study question – the sample case study on p. 25 (Lake District) could be used to answer this, but you should use a glaciated area you have studied.

a For this question the sketch map should be specific as it asks not just for the location but for named features within it. It is more likely to be an area of glacial erosion than deposition, as the features and evidence are easier to see in the landscape.

b You must base your answers / list on the features you have drawn / labelled on your map for (a).

c This is an important type of question as it looks at some of the issues involved when different types of human activity cause conflict. Typically this involves local residents, tourists / visitors and landowners or people working in the area
e.g. plans to dam an attractive glacial valley in order to build a reservoir to supply people with water.
THINK: what causes the conflict and between whom? You are NOT asked for solutions – so don't give any!

Revision Unit 4: Coasts

The Check up answers, p. 27

1 fetch; **2** destructive; **3** constructive; **4** swash, backwash; **5** transportation, deposition; **6** longshore drift; **7** angle; **8** attrition; **9** dissolving; **10** hard, resistant; **11** more quickly; **12** notch, collapse; **13** wave-cut platform; **14** A = resistant (harder) rock, B = less resistant (softer) rock, C = bay, D = headland, E = erodes more quickly forming a bay, F = erodes slowly leaving a headland; **15** stack; **16** bar; **17** groynes, transported; **18** B, A, C, B; **19** energy; **20** revetments

Exam Practice answers, p. 32

Question A

a THINK: 'annotated' means labels which explain the process – in this case longshore drift. Refer to p. 28 for help.

b i) Between each set of groynes (wooden 'fencing' built at right angles to the land, down the beach) beach material is being moved from one side, with sand and pebbles piling up on the other side in the direction in which longshore drift is happening. This is repeated along the beach between each pair of groynes – so one side has a small amount of beach material whilst the other has much more.
ii) Groynes are built to stop the beach material being eroded and transported along the coast. They are a fairly simple, cheap and effective method of doing this.

Question B

a Whenever you start a sketch from a photo, draw a frame first (the same size as the photo is helpful). Put in the main lines (lightly) first e.g. skyline, shape of main features. This is a sketch, not a simple line drawing so gradually add more detail and shading to give your sketch more depth and accuracy.

b X = arch, Y = stack, Z = cave.

c Refer to the information on p. 29. THINK: mention the processes which help create each feature (you don't have to explain them).

d The headland will continue to be eroded. Stumps will wear down, stacks become stumps, arches become stacks and so on. The headland / cliff will be eroded inland, threatening buildings, roads etc. and forcing people to evacuate homes.

Question C

This is a case study question – the sample case study on p. 31 (Mappleton) could be used to answer this, but you should use a coastal area you have studied. However, this is only for part (c) of the question – not all of it. THINK : always read the question very carefully so you know which part (s) you need to use a case study for.

a abrasion / corrasion, attrition, hydraulic action, corrosion: refer to p. 28

b sea walls, groynes, revetments, rock boulders, rock cages: refer to p. 30. If you choose to draw diagrams, annotate them carefully so that your annotations describe the method and how it works to reduce erosion.

c Case study part of the question. THINK: be careful to name the area. Explain clearly all the different methods being used to prevent or reduce erosion – use as much detail as you can in terms of names of places, figures or data.

Revision Unit 5: Weather and climate

The Check up answers, p. 33

1 atmosphere; **2** average, 25 / 30; **3** latitude, altitude; **4** temperature, precipitation; **5** highest, lowest; **6** west, prevailing; **7** Equator, away; **8** warmer, drier; **9** the sun's rays have to travel further and / or the same amount of energy has to heat a larger surface area nearer the Poles; **10** precipitation, wind, pressure; **11** anemometer; **12** 12°C, 13–17 knots, south / south west, drizzle, 4 oktas; **13** condenses; **14** relief (orographic), frontal; **15** warm / hot, dry;
16 (See diagram below);

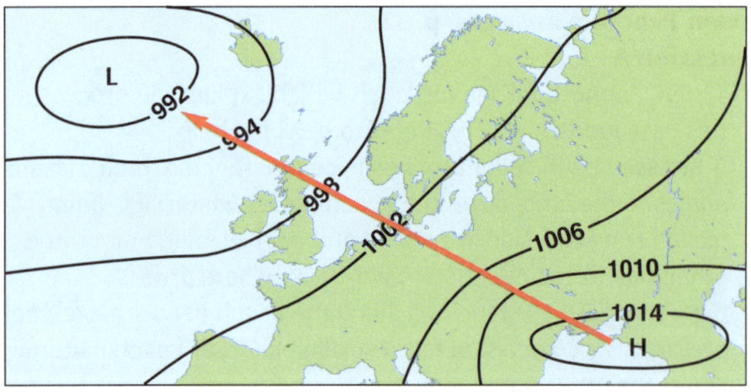

17 depressions; **18** warm, cold; **19** rises; **20** hurricanes, typhoons, cyclones

Exam Practice answers, p. 38

Question A

a = 2–3°C, b = approx. 1800 mm,

c temperatures are hot and fairly constant all year, around 25°C. Precipitation is high but variable. The wettest months are from December to May, with only half that amount falling from June to November.

d This is a tropical climate. There are constant high temperatures all year with heavy rainfall.

Question B

a an area of low pressure

b (See diagram)

c i) As the warm front approaches it becomes warmer and starts to rain, and air pressure falls. As the warm front passes over, the temperature continues to rise and the skies become clearer as the rain stops. As the end of the warm sector approaches, heavy rain falls, temperatures drop and air pressure falls. After the warm sector passes it is colder and showery. Refer to p. 35.
ii) This system is a low pressure system or depression.

Question C

This is a case study question – the sample case study on p. 37 (Hurricane Andrew) could be used to answer this, but you should use a weather hazard / event you have studied.

a This may be an extreme weather event e.g. very low pressure, heavy rain, no rain etc. Describe carefully the actual cause e.g. how is a hurricane caused, why did a drought happen. (No rainfall – why? Global warming?)

b Give the precise location(s) e.g. countries, regions. A labelled sketch map may help.

c Describe the different types of damage e.g. buildings damaged, power lines blown down. You may want to differentiate between short-term and long-term damage e.g. short term = roads blocked; long term = ruined crops affect the economy for several years.

d This refers to the cause of the damage itself e.g. flooding and high winds. Name the causes, indicating the broad type of damage caused (don't repeat the detail of answer(c)).

e Precautions can be to the area itself e.g. boarding up windows, using sandbags, or avoidance by e.g. evacuating people, moving them to safe areas.

Revision Unit 6: Ecosystems

The Check up answers, p. 39

1 living, non-living; **2** stores, outputs; **3** energy, sun; **4** producers; **5** biomass; **6** decomposers, nutrients; **7** A = leaves; B = nutrients, C = roots; **8** biomes, climatic; **9** rainforest; **10** bushes, grasses; **11** long, waxy; **12** plants, animals; **13** convectional; **14** emergents; **15** under, bare, lianas / vines; **16** humid, wet / damp; **17** A = surface, B = washed, C = shallower, D = flooding; **18** infertile; **19** carbon dioxide, warming; **20** medicines

Exam Practice answers, p. 44

Question A

a A = owl, B = mice, C = caterpillar, D = grasses

b i) flowers, ii) woodlice

c If all the caterpillars died fewer leaves would be eaten. Small birds would need to look for other food, and their numbers would be reduced. The effects of this would continue up the food chain.

d i) Biomass is the amount of living material in the ecosystem;
ii) The primary producers.

e Decomposers are vital to the food chain as they break down dead or waste matter and return it to the soil so that the nutrients aren't lost and can be used again.

Question B

i) A = acacia tree, B = baobab tree, C = thorn bush, D = grasses
ii) Two from e.g. leaves lost in dry season to conserve energy / water; long roots to reach water deep underground; wide trunk stores water; small leaves lose less water through transpiration.
iii) The large canopy provides shade from the very strong sun.
iv) Grasses die back, retaining their roots. When it rains, they can quickly grow again.

b THINK: if you draw a diagram you can base it on the one shown, adding labels explaining the changes which have happened. If not, describe in detail the results of deforestation and farming.

Question C

This is a case study question – the examples on pp. 41-2 (Tropical rainforests and savanna) could be used to answer this, but you should use an ecosystem you have studied.

a Name the type of ecosystem and the particular location you are writing about – a sketch map could be used. THINK: It MUST be an area which has undergone change.

b THINK: to describe changes you need to establish what was there BEFORE it changed – this is almost a comparison between then and now. Describe each of the changes in turn and how the changes have happened.

c Try to explain at least two, perhaps three, good effects and two, perhaps three, bad effects of the changes to the ecosystem. You must write about both to gain full marks.

Revision Unit 7: Population

The Check up answers, p. 45

1 evenly; **2** densely; **3** number, area; **4** double, 12; **5** Demographic, 4; **6** birth, babies, 1000; **7** birth, death; **8** Less Economically Developed, More Economically Developed; **9** birth; **10** health, living; **11** A = 5, B = economically, C = elderly; **12** Less Economically Developed, wide, narrow; **13** More Economically Developed, rectangular; **14** young, elderly; **15** movement; **16** seasonal, temporary; **17** rural, urban; **18** counter-; **19** push; **20** pull

Exam Practice answers, p. 51

Question A

a i) densely populated: many people in a given area (usually per km^2);
ii) sparsely populated: few people in a given area (e.g. per km^2).

b i) and ii) name any three for each – they may be counties or parts of counties.

c It is misleading because it is an average worked out by dividing the total number of people into the total size of the country. Many areas will have a higher density, many will have a lower density.

d THINK: when you describe distribution you are looking at where something is found. You can be fairly general with your answers e.g. most people live in the south east (around London), South Wales (around Cardiff), the Midlands (around Birmingham), the north west (Manchester and Liverpool), Yorkshire (Leeds and Sheffield), the north east (Newcastle) and the lowland areas of Scotland. Sparsely populated areas are north-west and southern Scotland, mid and North Wales, Cumbria and parts of the south west.

e Part (d) was just a description – here you explain why there are differences in density e.g. most of the sparsely populated areas are in the highland areas where it may be cold and wet; most of the densely populated regions are in lowland areas where it is warmer and drier and easier to build.

Question B

a i) approx. 50%, ii) approx. 5%

b 15–65 years

c 15–19, 40–44

d A = triangular / pyramid shaped, B = roughly rectangular

e i) A = LEDC, B = MEDC
ii) LEDCs are classically triangle / pyramid shaped, with a wide base and narrow top. MEDCs are more rectangular in shape, with fairly similar proportions in most ages groups.
It's easier to draw this than write a description.

f You would see fewer people in the 40-60 age group, especially on the male side, as this group would have been the soldiers etc. involved in the war.

Question C

This is a case study question – the sample case studies on p. 50 (Mexico / USA migration) could be used to answer this, but you should use an example of migration you have studied.

a Make it clear from which area / country the migrants came, and where exactly they went to. A labelled sketch map could be used.

b Refer to p. 49 for some general ideas – look at your own notes for factors specific to your chosen case study.

c These may be connected with the actual migration itself, but more likely any problems encountered once they had settled in their new country e.g. discrimination from local people.

d It is mainly young men who migrate, often looking for work. They leave behind the young, the elderly and the women – this can create difficulties e.g. lack of farm workers, further decline in the economy. THINK: write about as many effects as you can, explaining how they came about.

Revision Unit 8: Settlement

The Check up answers, p. 52

1 site; **2** situation; **3** hill; **4** function; **5** hierarchy; **6** more; **7** linear; **8** more; **9** convenience; **10** A = corner shops, B = suburban shopping parade / centre, C = CBD / Central Business District / city centre; **11** regional / out-of-town; **12** nine; **13** models; **14** Central Business District (CBD); **15** A = Light industry / manufacturing; B = medium-class residential; C = high-class residential / housing; **16** younger / newer / less tall; **17** rural to urban; **18** shanty towns / squatter settlements; **19** water, electricity; **20** self-help

Exam Practice answers, p. 60

Question A

a Any two: e.g. golf course (264401); gardens (256393); footpath – walking (265405). THINK: you MUST give six-figure grid references for each to gain marks.

b i) Combwich = village ii) Huntstile = hamlet
iii) Yearsley Farm = isolated settlement.

c Take care – there are two parts to this question: 1 – the feature is Durleigh Reservoir; 2 – it may supply water to Bridgwater.

d Refer to p. 53: sketch maps can be quite difficult. Look at how the example in the revision unit is drawn. You need a rough outline shape of the town – drawn to about the same size (then you can add a scale). Add important transport routes – motorways, main roads and railways – and physical features e.g. rivers, lakes, hills. Other important features might be the CBD, bridges and public buildings. You then need to annotate your map, adding explanatory labels about its site and situation. Don't worry if you find this difficult at first, it is quite a hard skill to master … keep practising!

e What is a market town? (Refer to p. 54) THINK: about location, surrounding land use (farming?), transport routes / accessibility.

Question B

a A = low-class residential / old inner city, B = middle-class/ inter-war housing, C = high-class / suburbs.

b Land is expensive and space limited – so it is cost effective to build upwards.

c Generally they get newer, less tall and with more space around them.

d THINK: about what each model shows and how model 2 has changed from model 1. Remember not just to describe what the differences are but to explain them too! … Model 1 shows simple concentric rings, with oldest buildings in the middle. Model 2 takes into account transport routes e.g. roads, canals, rivers and railways since industry in particular will locate along such routes, radiating out from the centre of an urban area. It is a more realistic model.

e The main difference is that, in model 3, housing becomes of poorer quality as you move out from the CBD, with high densities on the edges. In 1 and 2, older less attractive housing is in the centre, improving as you move out, the outer zone being the most sparsely populated. (Just describe – you are not asked to explain.)

f Urban land use models explain more simply what can be quite complex in reality and allow us to compare and contrast different urban areas in terms of common characteristics.

Question C

This is a case study question – the sample case study on p. 58 (Rio de Janeiro, Brazil) could be used to answer this, but you should use a city you have studied.

Part 1

Photographs are quite common in exams – especially showing urban areas / quality of life. You are asked just to describe some of the things you see which could cause problems for those living here. Refer to p. 57 for help.

Part 2

a Don't forget to name the city and describe its location – a simple labelled sketch map could be drawn. Make sure you have chosen a city which IS in an LEDC, and where some attempt has been made to improve housing (so read the question very carefully).

b No explanations – what was the area like before improvements were made?

c Include as much detail as you can about what has been done to upgrade housing, and who has made the improvements.

Revision Unit 9: Economic activity

The Check up answers, p. 61

1 primary, tertiary; **2** A / B, Less Economically Developed / More Economically Developed, primary / tertiary; **3** capital, market, labour; **4** processes; **5** farming; **6** arable, animals; **7** nomadic; **8** irrigation; **9** plantation; **10** Green, Less Economically Developed; **11** Common Agricultural Policy (CAP) overproduction; **12** heavy, raw; **13** assembly; **14** TNCs; **15** globalization; **16** global, acid; **17** high-tech; **18** coal fields; **19** coastal, importing; **20** greenfield

Exam Practice answers, p. 70

Question A

a i) Boro = December-March, Aus = April-August, Aman = July-November.
ii) Boro = (when crop is sown) at 25-26°, increases gradually to 33°. Fairly dry – rain in March around harvest time. Aus = sown during hottest part of year (35-34°), harvested as it drops slightly to 31°; rainfall throughout, very heavy in July before August harvest. Aman = temperature fairly constant throughout (31°), slight drop to 29° when harvested in November. Very wet when planted, getting gradually drier towards harvest. THINK – part (ii) refers to growing season (planting / harvesting) so use the right terms.

b i) Intensive farming.
ii) Refer to p. 64 to help you.

c Small-scale farming to support the farmer and his family. THINK: The question refers to Bangladesh – so suggest animals / crops likely here e.g. rice, wheat, fruits and vegetables (bananas, limes, coconuts, aubergines), cattle.

d Hard work, small land area so limited production, little money made – can sell surplus if there is any. Vulnerable to natural events – flooding, drought; no contingency plans.

Question B

a Give a general answer, perhaps with a few specific states for each.
i) North-east USA e.g. Ohio, Michigan.
ii) South-west USA e.g. California, Texas.

b i) Refer to p. 65 for help
ii) THINK – describing it is quite straightforward. 'Account for' is another way of asking you to explain why . 'Describe' = the sunbelt states have shown great increases in population e.g. 66% in Nevada, 41% in Arizona. The rustbelt states have either increased very slowly e.g. Ohio, 2%, or actually decreased e.g. West Virginia, -7%. 'Account for' = many industries in the rustbelt states have declined. People have migrated to the south west where new industries, jobs and better weather are found – hence the difference in the rates of population change.
iii) e.g. The sunbelt states could become overcrowded, with pressure on housing and other services.

c This is a case study question – choose one carefully from those you have studied. THINK: where was the industry first located? How has it changed? Why has it changed?

Question C

This is a case study question – the sample case study on p. 69 (Toyota) could be used to answer this, but you should use a TNC you have studied.

a Give the TNC / company name and the country it is from – no further detail is needed.

b List the different products.

c Choose a location you have studied in detail e.g. Toyota in Derbyshire, UK.
i) Give reasons for the choice of site / location. THINK: market, labour, capital etc. Refer to p. 62.
ii) List the good and bad factors connected with the TNC operating in the country chosen in (c) (i).

Revision Unit 10: Resources and energy

The Check up answers, p. 71

1 renewable; **2** once; **3** renewable, sustainable; **4** energy; **5** fossil; **6** 25%, MEDCs, 75%; **7** indicator; **8** greenhouse, global, acid; **9** oil, gas; **10** uranium; **11** radioactivity, waste; **12** fuelwood; **13** renewable; **14** A = hydro-electric, B = reservoir, C = turbine; **15** turbines; **16** solar; **17** geothermal; **18** plant, animal; **19** energy / electricity, damage; **20** oil

Exam Practice answers, p. 76

Question A

a i) 91%.
ii) Fossil fuels, when burnt, produce greenhouse gases which can harm the environment and contribute to global warming.
iii) 1%.

b i) and (ii) refer to pp. 73-74 for help.

c Production = 1611 million tonnes (equivalent); Consumption = 1952 million tonnes (equivalent).

d i) Oil and gas.
ii) This means the USA is not self-sufficient in these energy sources and has to import them.

Question B

a Hydro-electric power (producing electricity from water).

b i) Constant supply of water, water storage area, steep drop to channel water to turbines.
ii) Water forces the blades of a turbine round, which in turn works a generator to produce electricity – called hydro-electric power.

c Refer to p. 74 for help. (i) Your description or annotations should be detailed, clearly describing how this source produces energy. (ii) THINK: advantages and disadvantages compared with e.g. non-renewable sources / fossil fuels.

Question C

This is a case study question – the sample case study on p. 75 (Alaskan oil / *Exxon Valdez*) could be used to answer this, but you should use an example of an energy-producing region you have studied.

a Little detail needed – a map could be used.

b THINK: what did the region have before? How has this improved e.g. the economy, the lives of local people?

c Describe the type of damage which has occurred – either constantly or via a specific event. How has this happened?

d This could be long-term – to stop damage happening – or short-term to deal with damage when it does occur.

Revision Unit 11: Development

The Check up answers, p. 77

1 More Economically Developed Countries, Less Economically Developed Countries; **2** south; **3** Less Economically Developed Countries (LEDCs); **4** primary; **5** indicators; **6** low, raw; **7** B, C, A; **8** life expectancy; **9** first; **10** low; **11** calories; **12** Gross National Product, $US; **13** Human Development Index; **14** life expectancy; **15** countries; **16** measured; **17** five; **18** five, tertiary; **19** short; **20** Non-Governmental Organisations

Exam Practice answers, p. 83

Question A

a i) Japan, USA, UK.
ii) THINK: GNP is the most common way of judging economic development.

b THINK: There are many indicators used to measure development, and they cover a great range of values. Some countries are in the process of improving their economies, and are not really either LEDCs or MEDCs (they are sometimes referred to as NICs – Newly Industrialized Countries).

c These factors are mainly a good diet and good health facilities – controlled by the level of wealth. Explain why these affect life expectancy.

d Refer to pp. 78-80.

Question B

Part 1

a i) Wealth. (ii) The country's earnings in a year divided by the number of people, given as $US per capita.

b i) and (ii) = two graphs.

c Graph (i) – positive correlation;

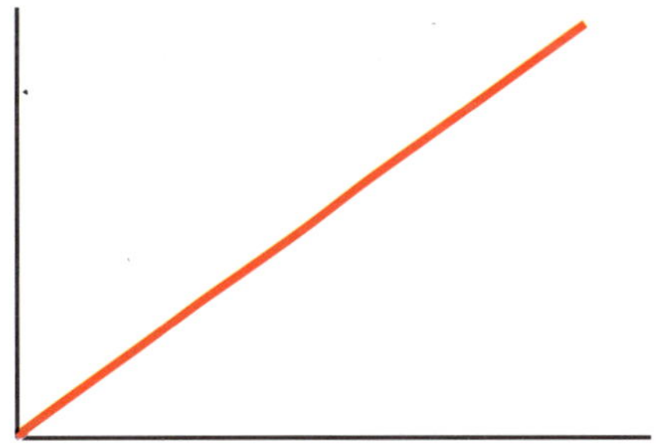

Graph (ii) – negative correlation.

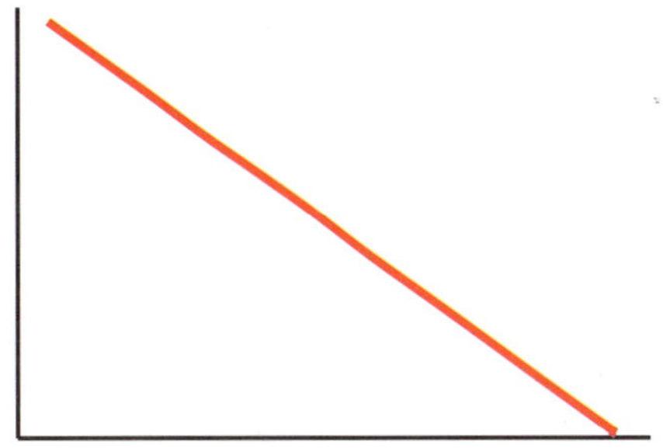

Part 2

a Refer to p. 80.
i) By using data for life expectancy, adult literacy / schooling and real GNP per capita ($)
ii) It does not just rely on wealth but uses social data as well.

b i) and ii) = map (various)

c The countries with the lowest HDIs are mainly in the south, especially Africa and Asia.

Question C

This is a case study question – the sample case study on p. 82 (regions of Brazil) could be used to answer this, but you should use contrasting regions in a country you have studied.

a This can be quite a simple map – country outline with two regions marked on and labelled.
b THINK: lots of factual detail needed here – precise data / information about the two areas will gain good marks – don't be too general e.g. more industry / less industry ... what industry?
c THINK: physical reasons, human reasons.

Revision Unit 12: Leisure and tourism

The Check up answers, p. 84

1 fewer / less / shorter; **2** leisure; **3** tertiary, service; **4** package; **5** domestic; **6** country; **7** independent; **8** long-haul; **9** fastest; **10** approx. 200 million; **11** environment; **12** low, temporary / seasonal / part-time; **13** foreign; **14** conflict; **15** infrastructure; **16** honeypot; **17** overcrowding, congestion; **18** footpaths; **19** ecotourism, environment; **20** levies

Exam Practice answers, p. 89

Question A

Part 1

a Honeypot sites.
b THINK: overcrowding – traffic, pollution, damage from too many people.

Part 2

a They are not nationally owned but owned privately by individuals, organizations or companies.
b THINK: what would different groups want or need – why would this cause disagreement or conflict?

Question B

a i) They are a long way from the UK; they are mainly in LEDCs.
 ii) Long-haul tourism.
b Refer to p. 85. THINK: increased leisure time, more money, cheaper travel.
c The question asks for benefits only. Refer to p. 86 for help.

Question C

This is a case study question – the sample case study on p. 88 could be used to answer this, but you should use a tourist destination / location you have studied.

a i) Remember to give both a general and specific location.
b THINK: what are the natural (and other) attractions?
c 'Affected' can mean for good or bad – deal with both where applicable for each of the three factors, (i), (ii) and (iii).

NOTES

This edition published exclusively for WHSmith, 2000, by:
Stanley Thornes (Publishers) Ltd

Reprinted in 2001 by:
Nelson Thornes Ltd
Delta Place
27 Bath Road
CHELTENHAM
GL53 7TH
United Kingdom

01 02 03 04/10 9 8 7 6 5 4 3 2

A catalogue record for this book is available from the British Library.

ISBN 0-7487-5491-1

Designed by Rocket Design Ltd, London

Produced by AMR Ltd, Bramley, Hants
Illustrations by Art Construction and Chartwell Illustrators

Printed and bound in Italy by G. Canale & C.S.p.A., Borgaro T.se, Turin

Acknowledgements

The authors and publishers are grateful to the following for permission to reproduce photographs and other copyright material in this book.

Andes Press Agency 57; Cumbria Picture Library 89; Kyodo News, Japan 11; Landform Slides 30; The MetroCentre, Gateshead, Tyne and Wear 59; Toyota Motor Manufacturing (UK) Ltd 69; Simon Warner 87 (top); World Pictures 32;

All other photographs are from the archives of Nelson Thornes Ltd.

The map extract and key on pages 92 and 93 are reproduced from the 1998 Ordnance Survey map of Weston-super-Mare (Landranger 182) reproduced with the permission of The Controller of Her Majesty's Stationery Office, © Crown copyright.

The authors and publishers also wish to thank David Waugh and Tony Bushell for permission to base diagrams in this book on certain diagrams from the Key Geography series.

Every effort has been made to contact copyright holders and we apologise if any have been overlooked.